MW00777157

Jon Tetsuro Sumida

Decoding Clausewitz

A New Approach to On War

University Press of Kansas

Published by the University Press of Kansas (Lawrence, Kansas 66045), which was
organized by the Kansas Board of Regents and is operated and funded by
Emporia State University, Fort Hays State University, Kansas State University,
Pittsburg State University, the University of Kansas, and Wichita State University

Library of Congress Cataloging-in-Publication Data
Sumida, Jon Tetsuro, 1949–
  Decoding Clausewitz : a new approach to On war / Jon Tetsuro Sumida.
    p.   cm. — (Modern war studies)
Includes bibliographical references and index.
ISBN 978–0–7006–1616–9 (cloth : alk. paper)
  1. Clausewitz, Carl von, 1780–1831. Vom Kriege.
  2. War.
  3. Strategy.
  4. Military art and science.
  I. Title.
  U102.S8415 2008
  355.02—dc22   2008012307

British Library Cataloguing-in-Publication Data is available.

Printed in the United States of America

10 9 8 7 6 5 4 3 2 1

The paper used in this publication is recycled and contains 50 percent postconsumer
waste. It is acid free and meets the minimum requirements of the American National
Standard for Permanence of Paper for Printed Library Materials Z39.48–1992.

442 R.C.T.

100 Bn.

M.I.S.

U.S. Army, 1942–1945

All nature is but art unknown to thee,
All chance, direction which thou canst not see.
    Alexander Pope, The Essay on Man

Many of the truths will make themselves felt here
only when one sees the chain connecting them with others.
    Charles de Secondat, Baron de Montesquieu, The Spirit of Laws

For although in a certain sense and for light-minded
persons non-existent things can be more easily and irresponsibly
represented in words than existing things, for the serious and
conscientious historian it is just the reverse. Nothing is harder, yet
nothing is more necessary, than to speak of certain things whose
existence is neither demonstrable nor probable. The very fact that
serious and conscientious men treat them as existing things brings
them a step closer to existence and to the possibility of being born.
    Albertus Secundus (Herman Hesse), The Glass Bead Game

Contents

Preface and Acknowledgments   *xi*

Introduction   *1*

1 Theorists   *9*
    Dismissal: Antoine Henri Jomini   *10*
    Advocacy: Julian Corbett   *17*
    Repudiation: Basil Liddell Hart   *25*

2 Scholars   *36*
    Text: Raymond Aron   *37*
    Context: Peter Paret   *50*
    Method: W. B. Gallie   *64*

3 Antecedents and Anticipations   *78*
    Historical Analysis   *80*
    Philosophical Invention   *94*
    Scientific Perspective   *112*

4 Imagining High Command and Defining Strategic Choice   *121*
    Absolute War and Genius   *121*
    History and Theory   *135*
    Defense and Attack   *153*

Conclusions   *176*

Appendix One. A Pictorial Representation of Critical Analysis   *195*

Appendix Two. Bach's *St. Matthew Passion* as a Model
for Critical Analysis   *197*

Notes   *199*

Bibliography   *213*

Index   *225*

Preface and Acknowledgments

G ENERAL HELMUTH VON MOLTKE, the architect of victory in the major conflicts that unified Germany between 1864 and 1871, regarded Clausewitz's *On War* as comparable in importance to *The Iliad* and the Bible.[1] In 1911, Julian Corbett, Britain's preeminent naval strategic theorist, observed that *On War* was "more firmly established than ever as the necessary basis of all strategical thought."[2] The distinguished German military and naval historian Herbert Rosinski maintained more than half a century later, in 1966, that Clausewitz's masterpiece was "the most profound, comprehensive, and systematic examination of war and its conduct" in existence, and then declared, "It towers above the rest of military and naval literature, penetrating into regions no other military thinker has ever approached."[3] In 1976, Bernard Brodie, America's pioneering theorist of nuclear strategy, argued that *On War* was "not simply the greatest but the only truly great book on war."[4] Clausewitz's intellectual biographer, Peter Paret, was able to state with confidence in 1986 that *On War* was "read more widely" in the late twentieth century "than at any time since it was first published in the 1830s."[5] Colin Powell, America's distinguished soldier-statesman, observed in 1995 that Clausewitz's work was "a beam of light from the past, still illuminating present-day military quandaries."[6] And in 2006, the "Professional Reading List" of the U.S. Army Chief of Staff proclaimed that *On War* was "one of the greatest works on military thought and strategy ever written"; it remained "essential reading for all senior leaders."[7]

Consumption of Clausewitz's work has been a phenomenon in the United States. Multiple English-language editions are in print,[8] a flow of interpretive books seems to be in progress,[9] and a dedicated website not only provides biographical information about Clausewitz but also a comprehensive guide to the critical literature and whole texts of writing by Clausewitz and Clausewitz scholars — it even sells Clausewitz posters, coffee mugs, T-shirts, and other trinkets.[10] Much of the interest in Clausewitz is generated by the American armed forces, which have made familiarity with *On War* an important feature of officer education. Reading at least some parts of the book is required for officers of every service. Exposure to Clausewitz occurs at each institutional stage of officer general education — at entry level in the service academies, at mid-level at the service command and staff colleges, and at the advanced level at the service and joint war colleges. No other book is regarded so highly. For the American military, *On War* is virtually a sacred text.

On War nonetheless poses great difficulties for readers. It is long, complicated, apparently inconsistent, in places seemingly obscure, and, to a large degree, concerned with military issues of the early nineteenth century that appear to be of little direct relevance to present concerns. In addition, many believe *On War* to be severely incomplete, raising the possibility that the text is not a coherent statement. Scholarly investigation has clarified a number of problems, but much remains enigmatic. As a consequence, there is no consensus with respect to *On War*'s general meaning, or even whether it can be said to have a general meaning. Given this state of affairs, the teaching of Clausewitzian thought has by default been based upon selective engagement.

In the simplest variant of selective engagement, the 500-plus pages of *On War* are reduced to a single phrase from the opening chapter: "War is merely a continuation of policy by other means."[11] This proposition is then interpreted as follows: Military action should be directed toward the accomplishment of policy (also political) objectives,[12] which ideally should be achievable. Hence the adjective "Clausewitzian" often refers to an attitude that recognizes the priority of political/policy considerations over military ones. The subject of the relationship of military action and politics/policy is especially important for democracies, where civilians and not the military bear the main responsibility for the formulation of political/policy goals. Clausewitz's dictum is thus a useful point of departure for an issue that has been and continues to be of great significance in the United States.

A less compressed form of selective engagement—which provides, therefore, a broader basis for general interpretation—views Clausewitz's first chapter as a précis of the entire book. In addition to the famous dictum on war and politics/policy, readers consider, among other things, his statements about what he called "absolute" and "real" war, the critical role of chance, and the notion that war is definable in terms of the relationships among emotion, courage and talent, and reason. The extension of analytical perspective from one sentence to a chapter permits application of *On War* to such questions as nuclear and conventional war, complexity and chaos theory, and the interplay among various human dimensions in strategic decision-making. These are all matters of considerable interest to the American military.

A third form of selective engagement is to cull significant observations and arguments from the entire book, which in effect adds concepts put forward in later sections to those described in the first chapter. They would include ideas such as friction, center of gravity, and culminating point of victory. A collection of this kind can be used in two different ways. In the first place, fragments of *On War* can be deployed piecemeal and opportunistically, without any attempt to identify theoretical organization. The book then becomes, in the words of Raymond Aron, the famous French social scientist, "a treasury of quotations."[13] Second, those Clausewitzian fragments that are judged to be germane to modern concerns can be assembled into an explanatory system. An analyst taking this approach simply discards the portions of *On War* that seem obsolete to him and reconstitutes the supposedly relevant remainder into what he believes to be a modernized and integrated whole.[14] The disadvantage of the first approach is that it begs the question of whether *On War* is informed by a coherent theoretical perspective. The disadvantage of the second is that it risks distortion or even contradiction of Clausewitz's original theoretical intent.

Selective engagement with *On War* has provided a basis for productive discussion of important military questions.[15] But the various and in some cases dissonant interpretations of Clausewitz's writing generated by selective engagement has prompted some to suggest that there is no general meaning to be found, and that the instruction offered is little more than platitudinous and even for the most part imagined. Azar Gat, a professor of political science and the author of a major history of strategic thought published in 1989, argues that "much of Clausewitz's reputation as a profound thinker has . . . resulted from the confusion among his interpreters." Thus, Gat observes, "Clausewitz could never have been wrong or less than profound because no one could be quite sure that he understood the true meaning of Clausewitz's ideas."[16] Bruce Fleming, a professor of English at the United States Naval Academy, maintains that "the range of possible interpretations of Clausewitz's messy [work] may well be the reason for its ongoing fascination: to a large degree it is a mirror of the person reading it." Thus he recommends "that *On War* be taught as poetry, even in the staff colleges, as an expression of the intrinsic contradictions of the human condition."[17] Tony Corn, an instructor at the U.S. Foreign Service Institute, remarks that *On War* is "ideally suited for endless, medieval-like scholastic disputatio," but is otherwise of little practical utility.[18]

While the length and difficulty of *On War* account for much of the propensity to resort to selective engagement, it is also the case that Clausewitz seems to have provided a measure of justification for viewing what he himself regards as an incomplete book in terms of a single sentence or just the first chapter. In a note of 10 July 1827 written in anticipation of the completion of *On War,* Clausewitz describes his book as "merely . . . a rather formless mass that must be thoroughly reworked once more." He subsequently declares that *"war is nothing but the continuation of policy with other means.* If this is firmly kept in mind throughout it will greatly facilitate the study of the subject and the whole will be easier to analyze."[19] In an undated note that some analysts believe was written in 1830, Clausewitz confesses that he was "dissatisfied" with "most" of the manuscript, and that he would have to "rewrite it entirely." He later writes that "the first chapter of Book One[20] alone I regard as finished," advising his reader that it "will at least serve the whole by indicating the direction I meant to follow everywhere."[21]

These memoranda—and in particular the note supposedly produced in 1830, the year before Clausewitz died—have promoted the view that Clausewitz had, on account of the unfinished state of the manuscript, sanctioned certain forms of selective engagement. This presumption has been the basis of rationalization—when rationalization is offered—of extreme reductionist representations of Clausewitzian thought. The evidence that Clausewitz himself had advocated reading only a small part of *On War* or interpreting the whole in terms of a part, however, is problematical for two reasons.

In the first place, there are the implications of certain statements made in the note of 10 July 1827. In this document, Clausewitz insisted that the imperfect condition of his book notwithstanding, a reading of the first six books as they were would enable "an unprejudiced reader in search of truth and understanding" to discover "basic ideas that might bring about a revolution in the theory of war."[22] It is clear, therefore, that Clausewitz believed that arguments challenging existing conceptions of war in a fundamental way could be discerned from a careful reading of the entirety of his unrevised text, and that, whatever these arguments were, they did not require exposition from the expected books on attack or war plans (what became Books VII and VIII).

In the second place, Azar Gat, in a book first published in 1989, made a strong case for dating the composition of the undated note to 1827 or even earlier. Gat offered five arguments in support of this view. First, the supposition that the note was written in 1830 was based upon nothing more than Marie von Clausewitz's observation that it had apparently been produced at a late date. Second, the manner in which Clausewitz discussed a proposed Book VIII suggests that the date of composition of the undated memorandum was before that of the note of 10 July 1827. Third, the note of 1830 refers to Books VI and VII as being no more than sketches or outlines, whereas the known texts of these sections are well-drawn and seem fully developed—Book VI is three times the length of most of the other books and discusses matters characteristic of Clausewitz's later thought, namely the relationship between war and politics. Fourth, the undated note does not mention or allude to the relationship between war and politics, or to absolute and real wars, ideas that Gat contends Clausewitz would have addressed in any description of the state of his thinking in 1830. And fifth, Marie von Clausewitz wrote that her husband believed in November 1831—the month of his death—that he could finish the manuscript of *On War* over the course of the winter, which Gat not unreasonably believes is an indication that the manuscript was in an advanced state of composition.[23] If Gat is correct about the earlier dating of the note supposedly written in 1830, this would allow consideration of the possibility that Clausewitz revised his manuscript between 1827 and 1830, which was a long enough time to make substantial changes and, indeed, to bring the book to near completion, if not completion.

There is, in light of the foregoing, very good reason to hold that at the very least the first six books of *On War* are capable of enabling a reader to come to terms with what their author believed was most important, and that probably all eight constitute an essentially sound representation of Clausewitz's considered views. This being so, there is no justification for either restricting careful reading to the first chapter of Book I or imposing its supposed perspective on the balance of the text.

In the opening paragraph of the first chapter of the first book of *On War*, Clausewitz informs the reader that he will begin with an examination of the "nature of the whole" of his subject and then consider the elements and parts of war and finally its internal structure. This brief outline probably refers to the order of books. The consideration of the "nature of the whole" appears to be the concern of Books I and II, entitled "On the Nature of War" and "On the Theory of War," respectively. The elements of war seem to be strategy, tactics, and the operative characteristics of armies, which are covered by Books III through V, entitled, "On Strategy in General," "The Engagement," and "Military Forces." The parts of war are almost certainly defense and attack, which correspond to the titles of Books VI and VII, respectively. The internal structure of war is provided by Book VIII, which is called "War Plans."[24]

Book I is divided into three parts. Chapters 1 through 3 are, respectively, about the complexity and changeableness of war, the effect of these characteristics upon the means and purpose of war, and the need for high intelligence and xvi

Preface and Acknowledgments

even-tempered yet energetic leadership on the part of the commander-in-chief to make decisions in the face of the difficulties of war arising out of these realities. Chapters 4 through 7 focus attention upon certain situational factors inherent to the conduct of war, collectively called "general friction," that obstruct or otherwise undermine the effectiveness of decision-making on the part of the commander-in-chief. Clausewitz believes that a commander-in-chief's intrinsic talent with respect to intelligence and executive capacity are not enough to surmount the problems posed by general friction. In Chapter 8, he thus maintains that a third variable, namely experience, is required, and that existing substitutes for experience are inadequate. In doing so he implicitly poses the question of what might constitute an effective palliative for lack of experience.

In Book II, Clausewitz provides his answer in the form of a discussion of the proper uses of and relationship between history and theory. Books I and II in combination lay the foundation for two major arguments about what can be learned from the study of history. The first argument is that an imagined replication of past decision-making of a commander-in-chief—that is, historical reenactment—can promote the development of a sensibility similar to that produced by the experience of real war. The second argument is that a sound review of history proves an essential truth about the governing dynamics of armed conflict, namely that the defense is a stronger form of war than the attack.

Books III through V, Clausewitz's treatise on the elements of war, address aspects of the first argument. Books VI and VII, on the parts of war, constitute his exposition of the second. The examination of the internal structure of war in Book VIII relates the interdependent elements and parts of war to the ends of war—that is, it explores the connections among strategy, tactics, and the nature of armies; defense and attack; and the political nature of armed conflict. A chapter on the structure of supreme command in this book was planned but never written.[25] Clausewitz had been motivated to write *On War* in large part to address the danger that he believed was posed by a politically revolutionary and militarily resurgent France to Prussia's national existence. He therefore concludes his last book with a description of the specific planning requirements of such a conflict.

In this book I consider the entirety of *On War* but focus my analytical attention on elucidating Clausewitz's primary argument. As a consequence, I examine certain books more closely than others. Clausewitz explained his theoretical approach in general terms in Books I, II, VI, VII, and VIII of *On War,* and the specific aspects of its application in Books III, IV, and V. Although I have carefully studied the books dealing with application, my main emphasis here is Clausewitz's general method. Thus I have concentrated my analysis on the first group of books listed above. For Clausewitz, the most challenging aspects of decision-making by the supreme commander were those concerned with strategy. He thought that tactics posed less taxing and certainly less important problems. Clausewitz's greater concern with strategy is evident in the text of *On War*, and that emphasis is reflected in the analytical balance of this book.

Four factors make comprehension of *On War* virtually impossible for most readers: faulty preconceptions, ignorance of historical context, the sophistication of Clausewitz's philosophical reasoning, and the complexity and difficulty of his exposition. For these reasons, Clausewitz's great work has the characteristics of an enciphered text, which must, as a consequence, be "decoded." In the present book, my method of accomplishing this task has been to use the contents of *On War* and other materials to elucidate the distinctive features of an intellectual persona. This knowledge was then deployed to make sense of otherwise inaccessible writing. My study is not, therefore, so much a book about a book as a book about a remarkable mind and how that mind can be engaged. I believe that taking this course has enabled me to clarify much that has mystified general readers, military professionals, and even expert scholars. It is also necessary to state, however, that my intended objective is neither to produce an executive summary of a text nor to devise the definitive interpretation of that text, but to orient the learning perspective of a reader of *On War* in a manner that will enable that person to negotiate this book as its author intended and, en passant, other things that matter a good deal as well.

My serious scrutiny of *On War* began in 1992 as I started to explore its role in professional military education in discussions with faculty at the U.S. National War College. I tested ideas and methods in my upper-division undergraduate seminar on Clausewitz, which I have taught annually at the University of Maryland, College Park, from 1993 onward, and in my Clausewitz seminar at the U.S. Marine Corps School of Advanced Warfighting, which I have offered each year since 2003. My thoughts were further developed during a semester of teaching in the Department of Military Strategy and Operations of the U.S. National War College in the fall of 2000, and in occasional lectures and seminars at the U.S. Army Military History Institute, the U.S. Army Advanced Strategic Arts Program, the U.S. Air Force School of Advanced Airpower Studies, the U.S. Marine Corps Command and Staff College, the Australian War Memorial, the Australian Defense Force Academy, Kings College (London), Yale University,

Preface and Acknowledgments

Oxford University, and the annual meeting of the Society for Military History in 2004. I completed most of the research and writing for this book between July 2004 and June 2006 at the U.S. Marine Corps University during my tenure as the Major General Matthew C. Horner Chair of Military Theory. The present work supersedes my published preliminary findings.[26]

Over the past fifteen years, I have benefited from discussion and correspondence with the following colleagues and friends: George Baer, Joseph Brinely, Ted Bromund, Jeffrey Clarke, Alexander Cochran, Michael Davis, Richard DiNardo, Jerry Driscoll, Arthur Eckstein, John Flach, John Gaddis, James Goldrick, Andrew Gordon, Michael Handel, Andreas Herberg-Rothe, Paul Kennedy, Jennie Kiesling, John Koenig, Nicholas Lambert, Emmet Larkin, William McNeill, Brad Meyer, Daniel Moran, Forrest Morgan, Christopher Owen, Rex Passion, Kevin Phillips, Mark Pizzo, David Rosenberg, Gordon Rudd, Hew Strachan, Joe Strange, Doug Streusand, Craig Swanson, David Syrett, Bruce Vandervort, Charles White, and Jonathan Winkler. My research and writing efforts were assisted greatly by the staff of the U.S. Marine Corps Research Center, Quantico, Virginia. I am especially indebted to Sir Michael Howard and Christopher Bassford, who read, engaged, and challenged my thoughts at every stage of my study of Clausewitz. The great bulk of the considerable translation effort involved in reviewing German-language sources was carried out by Ingo Trauschweizer, formerly a graduate student in the Department of History, University of Maryland, College Park. Final revision of the manuscript was greatly facilitated by the commentary of Robert Citino and Terence Holmes and by the intelligent support of my editor, Michael Briggs; the meticulous labor of my copyeditor, Kathy Streckfus; and the enthusiastic professionalism of the editorial staff at the University Press of Kansas.

Thomas A. Saunders III, Mary Jordan Horner Saunders, and the Marine Corps University Foundation financed my two consecutive years of academic leave through their sponsorship of the Major General Matthew C. Horner Chair of Military Theory at the U.S. Marine Corps University, Quantico, Virginia. I am deeply indebted to their generosity, which provided the main support for the critical late stages of my research and writing effort. I also owe considerable thanks for major funding from the Department of History, University of Maryland, College Park, which augmented the support given under the auspices of the Horner Chair, and to the International Security Studies Program at Yale University for a large lecture stipend. And finally, my parents, Mr. and Mrs. Theodore Sumida, provided essential financial assistance.

Preface and Acknowledgments

While I have been the recipient of a great deal of counsel from many individuals and considerable material support from a variety of sources, including those of national governments and related agents, the views expressed in this book are my own, and I assume full responsibility for them.

Jon Sumida

University of Maryland, College Park

November 2007

Decoding Clausewitz

Introduction

O*n War* IS WIDELY BELIEVED to be the greatest study of armed conflict ever written. This broad consensus among soldiers and scholars with respect to the authority of Carl von Clausewitz's magnum opus, however, is not accompanied by agreement about what it means. In the present work I address this unsatisfactory state of affairs in four ways. First, I evaluate a select body of important critical writing about Clausewitz and *On War*. Second, I describe the nature of Clausewitz's experience during the Napoleonic Wars and his intellectual reaction to that experience. Third, I consider Clausewitzian thought in terms of analytical perspectives offered by twentieth-century philosophical and scientific work. And fourth, I explain Clausewitz's practical and theoretical intent, his organization of subjects, the structure of his argument, the significance of his key terms, and his philosophical method. My findings challenge the interpretations found in earlier works on Clausewitz in multiple ways. But as Isaiah Berlin observed of Machiavelli, "where more than twenty interpretations hold the field, the addition of one more cannot be deemed an impertinence."[1]

A great deal of study has been devoted to *On War,* much of it intelligent, rigorous, and productive. Three forms of analytical misadventure, however, have prevented even these efforts from reaching satisfactory findings about fundamental issues. First, most critics have assumed that *On War* is unfinished to the point of requiring considerable compensatory interpretation. Second, the great majority have ignored Clausewitz's contention that the defense is a stronger form of war than the attack. And third, practically all have seen *On War* as an explanation of the general dynamics of armed conflict—that is to say, that it offers a theory of a phenomenon.

Those who maintain that Clausewitz did not believe he had completed *On War* have taken one of two positions: that his mature ideas on armed conflict were never presented in their final form, or that his considered opinion was expressed in some parts of the text but not in others. The effect of the former has been to open the door to speculation about where Clausewitz may have been heading, or denigration of what are supposedly immature and thus flawed ideas. The effect of the latter has been to promote analysis focusing only on those sections of *On War* that are thought to present his "late" views—a small portion of the whole. Disregarding Clausewitz's position on the relative merits of the defense and offense, however, amounts to the dismissal of what he believed was the most important strategic lesson of the great wars of his time. Moreover, those who view *On War* as a theory of a phenomenon have failed to see—or perhaps more correctly, have even failed to contemplate the possibility—that it might be something else.

My analytical perspectives differ from those just described in fundamental ways. First, I regard Clausewitz's text as sufficiently finished to express its author's main arguments clearly. This position is based mainly upon previously described analysis of memoranda written by Clausewitz about the manuscript that was to become *On War,* which enables me to set aside widely held views about its state of completion at the time of the author's death. Second, I devote a great deal of attention to Clausewitz's view that the defense was superior to the attack. This course is justified by Clausewitz's own declarations as to the importance of this subject and by the very large proportion of *On War* given over to the consideration of this question. And third, I examine *On War* as a theory of practice rather than as a theory of a phenomenon. A theory of practice, as the phrase implies, is about the *how* of learning to do something rather than the *what* of something's general nature. This is a complex argument that requires further explanation.

Clausewitz resorted to the invention of a theory of practice because he was convinced that no theory of war as a phenomenon was capable of representing the nature of war as it occurred in reality. He knew that any description of war in terms of general propositions could not account for the variety and in some cases improbability of circumstances that governed particular cases of armed conflict, and that this problem could not be solved or even mitigated significantly by statements that admitted exception to general propositions. This problem mattered because Clausewitz believed that a theoretical approach capable of fostering an understanding of armed conflict comparable to the experience of actual war could improve officer education and promote sound thinking about strategy. The attainment of such objectives was vital to Clausewitz, who was certain that Prussia's survival in the face of the dire threat posed to its continued existence by France depended upon the competence of the high command of the army and the national political leadership. His views were based on personal experience—in 1806, he had seen firsthand how military and political ineptitude could cause catastrophic defeat and the near destruction of national sovereignty.

Given the shortcomings of any set of general propositions as a representation of the reality of war, Clausewitz devised an alternative approach that focused on the decision-making capability of successful commanders-in-chief of the past. Such men, he argued, had demonstrated a practical understanding of armed conflict through their possession of whatever it took to overcome the manifold and in many cases unpredictable obstacles to taking sound action in war. He observed that a very considerable proportion of practical understanding was the product of experience. And he was furthermore convinced that the most important thing improved by experience was intuition, a form of unconscious thought that integrates rational intelligence and emotion.

Intuition was the basis of what Clausewitz called the "genius" of a commander-in-chief, and he believed that an understanding of this genius was central to any understanding of war's nature. The emotional component of intuition and the manner in which it could be combined with rationality could not be described in words, however, and thus in order for the quality of genius to be known, it had to be experienced, at least to some degree. Clausewitz's theory of war therefore offered instruction as to how a sense of intuition, and thus of genius, might be acquired by someone without experience of supreme command. This was to be done through the mental reenactment of historical case studies of command decision. And because many of the factors that formed the basis of command decision in reality were either intangible or, even if tangible, unverifiable from the historical record, reenactment of command decision required one to take into account factors whose existence and nature had to be surmised. A critical element of Clausewitzian theory therefore consists of descriptions of the dynamics of war as the basis of intelligent speculation that supported reenactment, and not as an end in itself. Clausewitz regarded such an indirect method of using general propositions as the best and indeed perhaps the only valid approach to using such instruments to achieve an understanding of war.

The effect of seeing the text of *On War* as a complete rendition of Clausewitz's considered major views, as a promoter of the idea that the defense was a stronger form of war than the attack, and as a theory of practice that could improve the effectiveness of the Prussian army, in particular, as well as provide the basis of a sound theory of armed conflict, is to change fundamentally the nature of the critical problem posed by his great book. This has required the development of analytical approaches that differ substantially from those of other scholarship in eight major areas.

Introduction

First, the assumption that *On War* is unfinished has discouraged consideration of Clausewitzian thought as a complete and coherent entity. This in turn has promoted tendencies to study only those propositions that are regarded as relevant to whatever the writer perceives to be important with respect to the military and political affairs of his or her own time. I depart from this common approach by treating the entire body of Clausewitzian argument as an interdependent whole that must be understood in terms of the Prussian officer's desire to address the military affairs of his day. This demands, in particular, recognition of the degree to which the form and substance of Clausewitz's writing in *On War* were shaped by his fear that his own country was threatened by invasion and permanent subjugation.

Second, nearly all previous investigations have characterized Clausewitz's argument that "war is an extension of politics by other means" as his most important contribution to the theory of armed conflict. I instead relate this proposition to qualifying ones that give the phrase a meaning that differs substantially from that normally ascribed to it. In particular, I subordinate the aphorism to larger primary arguments about historical reenactment and the superiority of the defense to the offense as a form of war.

Third, many critics have regarded Clausewitz's description of a "paradoxical trinity"—which consists of passionate feelings of the people, the courage and talent of the commander-in-chief, and the political reason of the government—as an independent statement about the dominant tendencies of war. I prefer to consider it a point of departure for Clausewitz's argument about historical reenactment and the relative merits of the offense and defense, rather than a fraught and expansive general proposition about armed conflict.

Fourth, while others have recognized that Clausewitz's concept of "genius" is of central importance to his theory of war, explanations of how and why this is the case have been overgeneralized and incomplete. I attempt to elucidate Clausewitz's idea of replicating the psychological experience of supreme command through historical reenactment to provide the foundation for deeper understandings of genius and its theoretical function.

Fifth, I maintain that imprecise and even completely faulty interpretations of the phrases "absolute war" and "real war," and their mistaken use as synonyms for "unlimited war" and "limited war," respectively, have generated a great deal of unproductive discourse. I aim to produce a theoretically coherent explication of these important but heretofore problematical terms by connecting Clausewitz's statements to his views on the comparative strengths of the defense and offense in general, and on guerrilla war in particular.

Sixth, Clausewitz's writing about resistance to occupation by armed civilians and by small and dispersed groups of regular troops—that is, guerrilla war—is usually given short shrift in general studies of *On War*. I relate guerrilla war to major concepts and to Clausewitz's primary argument, presenting his explanation of the topic as a critical component of his thought. In addition, I use the subject of guerrilla war to highlight the disagreement between Clausewitz and Antoine Henri Jomini, his most famous competitor in military theory, over the strategic efficacy of the principle of concentration of force, a matter of fundamental theoretical significance that has been largely, if not completely, ignored in critical studies of their work.

Seventh, previous writing about the philosophical character of *On War* has been concerned with the extent to which Clausewitz's thinking was influenced by the work of leading thinkers of the eighteenth and early nineteenth century. As a consequence, much—indeed, perhaps too much—has been made of Clausewitz's use of Kant's principle of division or Hegel's dialectical form of argument. I, in contrast, argue that Clausewitzian theory, by being a theory of practice rather than a theory of a phenomenon, embodies ideas that anticipated those of later philosophers and the findings of twentieth-century mathematics and cognitive science. Clausewitzian theory can thus be seen as a philosophically and scientifically creative achievement rather than a mere adaptation of certain well-known philosophical approaches of his day.

Eighth and finally, the standard characterization of *On War* as a theory of a phenomenon has caused critics to read it as presenting a certain fixed understanding about armed conflict. Recognition that *On War* is a theory of practice redirects us to see the work as a guidebook showing the reader how to achieve greater knowledge about his own conscious and unconscious intellectual and emotional faculties, using certain propositions about major fighting between nation-states as a means of achieving that end, rather than as a means of achieving the command of concepts about war per se. In contradistinction to all previous major studies, I thus regard *On War* as a *set of instructions on how to engage in serious learning of a highly personal nature* rather than an *impersonal representation of the totality of that which is to be learned*. Clausewitz's approach to theory may be seen not only in terms of how it might improve an individual's decision-making capacity in war and politics, but also in terms of how it might be a pedagogical model applicable to the development of the ability to do anything that is difficult, complex, contingent, and dangerous.

Introduction

The contention that *On War* is about how to learn in a certain way rather than about what constitutes proper conduct determined the form of the conclusions of this book. Clausewitz maintained that it is the unique genius of individuals—and not theoretical propositions—that solve real strategic problems. For this reason he would almost certainly have regarded the direct application of his arguments to actual situations to be problematical, if not altogether inappropriate. In my conclusions I have therefore attempted to answer questions that respect Clausewitz's fundamental attitudes about the practical strengths and limitations of theory: What did he say to the Prussians of his day, why have his arguments in *On War* been so hard to negotiate, and how have his main ideas fared with the passage of time?

To lay the groundwork for my own interpretation of Clausewitz's *On War,* I first summarize and examine the views of other major theorists and scholars. The first chapter, on theorists, considers the writing of Antoine Henri Jomini, Julian Corbett, and Basil Liddell Hart, all three of whom had strong views on Clausewitz. Because of their stature as leading thinkers about the nature of war, their views on Clausewitz have been widely disseminated and have exerted great influence. Indeed, they may well have had more to do with shaping the general opinion of Clausewitzian thought than direct reading of *On War.* The second chapter, on scholars, examines the writing on Clausewitz of Raymond Aron, Peter Paret, and W. B. Gallie. Aron's book on Clausewitz is regarded as a fundamentally important study of the main arguments of the text, and Paret's is thought to be the best account of the intellectual context of *On War.* Gallie's two articles are, in comparison, almost unknown, but can be described as the most philosophically well-informed approach to the ideas of *On War* ever written, their errors notwithstanding. The writings of these six thinkers encompass the great bulk of what is known, or thought to be known, about Clausewitz and his work, and because of them readers of Clausewitz often come to his work with preconceptions that can be an obstacle to comprehension of his actual text. I have therefore made it my first order of business to assess this literature.

In the third chapter, I show how our view of Clausewitz might be affected by our knowledge of the historical military context of his time and by ideas from the fields of philosophy, mathematics, and cognitive science. In the fourth and final chapter—with prejudice thus dismantled and understanding broadened— I present a systematic examination of large sections of *On War.* The exposition in this last chapter is the product of conventional direct methods of literary criticism. Or in other words, the interpretation of text is not overdetermined by the knowledge of what is supposed to be context, but is based primarily upon the plain meaning of words and justifiable inference prompted by such plain meaning. This is important because the findings of Chapter 4 are anticipated in the earlier chapters in order to provide the correct points of departure for the examination of the sources and context of his thought. In places prior to Chapter 4, consideration of certain issues—namely, in Chapter 1, the theoretical significance of battle, and in Chapter 3, Clausewitz's views on the limited capacity of language to convey meaning, as well as the importance of intuition—requires the discussion of the text of *On War*. These sections are not, however, essential to the presentation of the main lines of argument in the fourth chapter, which is analytically self-contained.

Critics of the standard English-language translation of *On War* have identified problem areas that merit discussion but have yet to offer a persuasive case that translation error has compromised the statement of Clausewitz's major arguments. As should be clear from the foregoing discussion, it is my contention that the main cause of our failure to understand Clausewitzian thought up to this point is faulty conceptualization and the consequent application of inadequate instruments of analysis. From this it follows that serious misunderstanding of *On War* in translation has little to do with imprecise, misleading, or otherwise incorrect rendering of the German language. This line of criticism has thus not been pursued. In addition, my review of previous critical writing on Clausewitz in Chapters 1 and 2 is intended to perform the specific functions described above, and makes no claim to being comprehensive. For a general survey and evaluation of the voluminous and growing literature on Clausewitz, readers are referred to the publications of Christopher Bassford, Beatrice Heuser, Hew Strachan, and Andreas Herberg-Rothe, as well as the Clausewitz "home page" on the Internet.[2]

My main sources include the standard English-language translation of *On War*, the scholarly German-language edition of *On War*, a large selection of Clausewitz's other writing in translation and in German, a selection of the biographical and critical literature in English and German, selected scholarly studies of the military history of the period of the French Revolution and empire, and a number of nineteenth- and twentieth-century works on philosophy, mathematics, and cognitive science. Material in German was negotiated with the assistance of a native German speaker trained at the doctoral level in modern military history who was thoroughly familiar with my views on Clausewitz. I have not subjected all the untranslated texts of Clausewitz's collected works to

Introduction

exhaustive examination, nor did I consider every extant major approach to the elucidation of the text of *On War*. That said, I am confident that my examination of his work is more than sufficient to provide a solid basis for my conclusions about Clausewitz's great work.

My main objective in this book was to make Clausewitz's thought accessible to general readers and military professionals alike in a form that does justice to its coherence, originality, and power to provoke insight. To this end I have ignored certain matters that have been the concern of many others in the belief that discussion of the arcane, ill-founded, or anachronistic has obscured Clausewitz's main lines of argument. In choosing to forgo a reexamination of these matters, I have acted in the same spirit as Gallie when he wrote, "I am not suggesting that Clausewitz must therefore remain an impenetrably obscure thinker, the reserve of a few learned and logically skilled specialists, who alone can separate out what is best in him from what is confused and fallacious. On the contrary, it would be truer to say that I want to rescue Clausewitz from the Clausewitzian specialists."[3]

"The truly awesome intellectuals in our history," observed Stephen Jay Gould, the distinguished naturalist,

> have not merely made discoveries; they have woven variegated, but firm, tapestries of comprehensive coverage. The tapestries have various fates: Most burn or unravel in the footsteps of time and the fires of later discovery. But their glory lies in their integrity as unified structures of great complexity and broad implications.
>
> Yet, in our harried world of sound bites and photo ops, we focus on anecdote rather than structure, and scholars are identified by items rather than by their precious tapestries.[4]

My aim has been not only to examine *On War* as a tapestry in Gould's terms, but to argue that its fabric has neither burned nor unraveled, and thus retains much of its original power to prompt important thought and action.

1

Theorists

O N THE EVENING OF 16 NOVEMBER 1831, Major General Carl Philipp Gottfried von Clausewitz of the Prussian army died in Breslau at the age of
fifty-one. At the time of his death, he was a senior inspector of artillery, a post he
had assumed in August 1830. In the intervening months, his bureaucratic work
had been interrupted when revolutions in France and the Russian sector of par-
titioned Poland disrupted European political tranquility. Clausewitz hoped that
he would serve in what he believed would soon be a major war against France,
which posed the greater danger. But in December 1830 he was appointed chief of
staff to Field Marshal Augustus Wilhelm Gneisenau (1760–1831), who had been
given command of the forces deployed to the Prussian sector of Poland. In the
event, the Russians were able to restore order in their zone, eliminating the need
for Prussian military involvement.

Clausewitz was disheartened by his consignment to a secondary theater, in-
activity, and the burden of witnessing the infirmities of the aging Gneisenau, his
close friend and patron. Gneisenau's death from cholera in August 1831, the in-
sulting lack of notice on the part of King Frederick William III (1770–1840) to his
demise, and Gneisenau's replacement by a courtier general who represented political and military values that he detested exacerbated Clausewitz's already serious depression. In the first week of November, following the conclusion of the
crisis in Poland, he returned to Breslau. Even reunion there with his beloved
wife was not enough to restore his mental equilibrium. According to his doc-
tors, Clausewitz's death from what was considered to be a mild case of cholera
was caused in large part by preexisting psychological and physical weakness.

Clausewitz had planned to spend the winter completing a study that, in
his words, "tested each conclusion against the actual history of war."[1] His original intention in 1818 had been "at all costs to avoid every commonplace,
everything obvious that has been stated a hundred times and is generally believed," in order to "write a book that would not be forgotten after two or
three years, and that possibly might be picked up more than once by those

who are interested in the subject."² His ambition, he declared in 1827, was to present basic ideas that "might bring about a revolution in the theory of war."³ After Clausewitz's death, one of his close friends proclaimed the prospect of the book's publication. Clausewitz's "ideas of the art of war," observed Count Carl von der Groeben in an obituary printed in the *Staatszeitung,* "were forged from the deepest of research and burning experience. They were concerned with the highest policy, were comprehensive, and as simple as they were practicable. The writings which he has left us will also show him to those who did not know him personally. . . . His noble bereaved widow will not keep them from those who follow him."⁴

Clausewitz's literary estate was in good order. Prior to assuming active service in the spring of 1830, he had, according to his wife, Marie von Clausewitz, "arranged his papers, sealed and labeled the individual packages." In addition, Marie was highly intelligent and thoroughly familiar with her late husband's thinking and manner of thinking on serious subjects. She was thus able, with the assistance of Clausewitz's friends and her brother, to convert the manuscript into print in only six months. Although the work involved "careful checking and sorting," the published volume, Clausewitz's literary executor insisted, reproduced her husband's original text "without one word being added or deleted."⁵ *On War* was published in 1832. Reviews were respectful and even admiring, although some commentators observed that the difficulty of the exposition would deter many readers. *On War* was not a popular success—twenty years after its release, a portion of the original print run of 1,500 copies was still available. It did, however, provoke a strong response from a rival author, who was widely considered to be the greatest military theorist of his generation.

Dismissal: Antoine Henri Jomini

Baron Antoine Henri Jomini (1779–1869) was born in Switzerland to a well-established family. In his youth, he supported proponents of the French Revolution in his native land, and he served in the Swiss government after the establishment of a radical regime. Following the rise of Napoleon, Jomini combined service as a soldier in the armies of France and Russia with historical scholarship. From 1805 to 1813 he was given a series of senior staff appointments in the French army, in the course of which he was awarded a barony and promoted to the rank of brigadier general. In 1813, he changed sides and joined the Russian army, where he was immediately given the rank of lieutenant general and made personal aide-de-camp to the Tsar Alexander I (1777–1825). Jomini was a gifted staff officer, but his fame was largely attributable to his prolific writing. His first book, *Treatise on Major Military Operations of the Seven Years' War,* was published in 1805. By 1811, he had extended this work of two volumes by four additional books to cover the first two years of the French Revolution. In 1813, when he left the French army, he was widely regarded as "the preeminent historian and theorist of modern warfare."[6] By the time of Clausewitz's death, Jomini had added eleven volumes to his account of war in the age of Napoleon (fifteen volumes, 1806–1824), produced a four-volume biography of Napoleon (1827), and combined his chapters and articles on military theory into a single volume entitled *Synoptic Analysis of the Art of War* (1830).[7]

In *On War*, however, Clausewitz assaulted Jomini's reputation with harsh criticism of his thought. There can be little doubt that Clausewitz is referring to Jomini, among others, when he writes that the recognition of an inherent gap between theory and practice defies "common sense" and constitutes nothing more than "a pretext by limited and ignorant minds to justify their congenital incompetence" (*On War*, Book II, Chap. 2, 142). Clausewitz is also probably including Jomini's work when he maintains that the "obscure, partially false, confused and arbitrary notions" of previous theorists has "made theory, from its beginnings, the very opposite of practice, and not infrequently the laughing stock of men whose military competence is beyond dispute" (ibid., II/5: 169). Clausewitz strikes at the heart of Jomini's literary achievement when he criticizes efforts to prove propositions through the study of multiple battles as amounting to "superficial, irresponsible handling of history" that results in "hundreds of wrong ideas and bogus theorizing" (ibid., II/6: 173). And Clausewitz is almost certainly aiming at Jomini when he declares that "one of the chief functions of a comprehensive theory of war" was to expose the "vagaries" and "fantasies" of "a favorite of recent theorists" (ibid., III/15: 215).

Jomini's response was to write what was to become his most famous book, *Summary of the Art of War, or, A New Analytical Compend of the Principal Combinations of Strategy, of Grand Tactics and of Military Policy,* which was published in 1838. In his opening bibliographical essay, "Notice of the Present Theory of War, and of Its Utility" (omitted in the standard translation), Jomini displays his annoyance with Clausewitz in no uncertain terms. He acknowledges that *On War* had "made a great sensation in Germany," but goes on to say that although one cannot deny to General Clausewitz great learning and a facile pen, this pen, at times a little vagrant, is above all too pretentious for a didactic discussion, the simplicity and clearness of which ought to be its first merit. Besides that, the author shows himself by far too skeptical in point of military science; his first volume is but a declamation against all theory of war, whilst the two succeeding volumes, full of theoretic maxims, proves that the author believes in the efficacy of his own doctrines, if he does not believe in those of others.

As for myself, I own that I have been able to find in this learned labyrinth but a small number of luminous ideas and remarkable articles; and far from having shared the skepticism of the author, no work would have contributed more than his to make me feel the necessity and utility of good theories, if I had ever been able to call them in question; it is important simply to agree well as to the limits which ought to be assigned them in order not to fall into a pedantry worse than ignorance. (Jomini, *Art of War* [Winship/McLean], 14–15)[8]

Clausewitz, according to the foregoing assessment, is guilty of being obscure, destructive of all scientific inquiry, yet hypocritical because he has articulated theories of his own that purport to be scientific, and short on useful conceptualization. "I have never," Jomini later observes, "soiled my pen by attacking personally studious men who devote themselves to science, and if I have not shared their dogmas, I have expressed as much with moderation and impartiality: it were to be desired that it should ever be thus" (ibid., 16). These serious charges do not, however, exhaust Jomini's list of protests.

Jomini ends the main text of his essay by lauding the collective achievement of other writers in advancing knowledge about war, including Clausewitz. But he could not resist adding an asterisk next to the Prussian author's name—the only one of the fourteen so treated—to direct readers to a highly critical, even defamatory, footnote. "The works of Clausewitz," Jomini there admits,

have been incontestably useful, although it is often less by the ideas of the author, than by the contrary ideas to which he gives birth. They would have been more useful still, if a pretentious and pedantic style did not frequently render them unintelligible. But if, as a didactic author, he has raised more doubts than he has discovered truths, as a critical historian, he has been an unscrupulous plagiarist, pillaging his predecessors, copying their reflections, and saying evil afterwards of their works, after having travestied them under other forms. (Ibid., 21)

Direct references to Clausewitz in the main body of the *Summary of the Art of War* are limited to two brief critical remarks. In the first, Jomini characterizes Clausewitz's logic as "frequently defective" (*Art of War* [Mendell/Craighill], 151).[9] as in the case of the Prussian author's alleged contention that maneuver by the defense in mountain warfare is to be avoided on the grounds that such action forfeits the advantages of local strengths, and mountainous terrain disfavors initiative on the part of the defender. What Clausewitz actually argues is that while fixed defenses in mountains can be held against superior numbers for short periods, which can be useful, they cannot prevent a breakthrough indefinitely. He also contends that mountainous terrain will impede movement on the part of the defense to counter the inevitable offensive breach, and that in the event of a mountain battle in which the main forces of the defender are defeated, the difficulty of retreat through mountains can result in the attacker achieving a decisive victory through effective pursuit (ibid., VI/15: 419–422; VII/2: 539). Clausewitz thus concludes that "mountains are generally unsuited to defensive warfare, from the point of view of both tactics and strategy," and that insofar as decisive battle is concerned, "mountainous terrain is no help to the defender; on the contrary, . . . it favors the attacker." These views are actually similar to the ones articulated by Jomini in the *Art of War*. Perhaps he misrepresented Clausewitz's arguments because he misread him, or perhaps he was simply annoyed at the Prussian general's contention that his views contradicted "general opinion," which, Clausewitz said, was "usually in a state of confusion and unable to distinguish between diverse aspects of a question" (ibid., VI/16: 423, 427).

In the second direct reference to Clausewitz, Jomini accuses the Prussian writer of having committed "a grave error" of military analysis when he supposedly maintained that "a battle not characterized by maneuver to turn the enemy cannot result in a complete victory" (*Art of War* [Mendell/Craighill], 163). The statement as given probably does not accurately represent Clausewitz's views. Jomini does not provide a reference, but he seems to be referring to the passage in *On War* stating that "a battle fought with parallel fronts and without an enveloping action is not so likely to bring great results as one in which the defeated army has been turned, or made to change its front to greater or lesser degree" (*On War*, IV/11: 261).[10] In any case, Clausewitz later makes clear that he is an agnostic when it comes to positive instructions on the subject of maneuver:

> We are therefore certain that no rules of any kind exist for maneuver,
> and no method or general principle can determine the value of the
> action; rather, superior application, precision, order, discipline, and

fear will find the means to achieve palpable advantage in the most singular and minute circumstances. It is on these qualities that victory in this type of contest largely depends. (Ibid., VII/13: 542)

Jomini also engages Clausewitz's writing indirectly—that is, without mentioning his name but clearly countering what he regards as his major arguments. Clausewitz dismisses the practical utility of rules and principles with respect to decision-making at the strategic level (ibid., II/2: 134, 136, 139–142; VI/30: 516–517). "All my works," Jomini writes defiantly in reply, "go to show the eternal influence of principles, and to demonstrate that operations to be successful must be applications of principles" (*Art of War* [Mendell/Craighill], 115). Any system of military practice, Jomini observes later on, that

is not in accordance with the principles of war cannot be good. I lay no claim to the creation of these principles, for they have always existed, and were applied by Caesar, Scipio, and the Consul Nero, as well as by Marlborough and Eugene; but I claim to have been the first to point them out, and to lay down the principal chances in their various applications. (Ibid., 116)

In his conclusion, Jomini writes:

If a few prejudiced military men, after reading this book and carefully studying the detailed and correct history of the campaigns of the great masters of the art of war, still contend that it has neither principles nor rules, I can only pity them, and reply, in the famous words of Frederick, that "a mule which had made twenty campaigns under Prince Eugene would not be a better tactician than at the beginning." (Ibid., 296)

Clausewitz argues that politics not only governs the decisions that lead to war, which he notes is generally understood, but also influences decision-making during the war (*On War*, I/1: 87; I/2: 92; VIII/6: 605–608). Jomini disagrees. While he acknowledges that there exists an "intimate connection between statesmanship and war" during the period *leading up* to hostilities, he warns that when "military enterprises are undertaken to carry out a political end" that is "sometimes quite important," the "irrational" character of the political objective will "frequently lead to the commission of great errors in strategy." For this reason, Jomini concludes, contra Clausewitz (ibid., I/1: 87,

92), that "political objective points should be subordinate to strategy, at least until after a great success has been attained" (*Art of War* [Mendell/Craighill], 82–83). Jomini uses this virtual separation of politics and strategy elsewhere to defend the validity of his view that there are fundamental principles of war of universal applicability, observing:

> If the principles of strategy are always the same, it is different with the political part of war, which is modified by the tone of communities, by localities, and by the characters of men at the head of states and armies. The fact of these modifications has been used to prove that war knows no rules. Military science rests upon principles that can never be safely violated in the presence of an active and skillful enemy, while the moral and political part of war presents these variations. Plans of operations are made as circumstances may demand: to execute these plans, the great principles of war must be observed. (Ibid., 15)[11]

Clausewitz argues that the defensive is a stronger form of war than the offensive. In particular, he maintains that a defender can trade space for time as the prelude to a counterattack delivered once the attacker has become overextended.[12] This idea is consistent with Napoleon's well-known maxim: "The whole art of war consists in a well-reasoned and extremely circumspect defensive, followed by rapid and audacious attack."[13] Jomini, therefore, had reason to favor the argument, which makes it one of the few ideas of Clausewitz that he could find useful. The "defensive-offensive," Jomini says,

> combines the advantages of both systems; for one who awaits his adversary upon a prepared field, with all his own resources in hand, surrounded by all the advantages of being on his own ground, can with hope of success take the initiative, and is fully able to judge when and where to strike. (*Art of War* [Mendell/Craighill], 67)

Guerrilla warfare—what Clausewitz calls "People's War"—however, exposes an important area of disagreement. Clausewitz had studied the insurrection of armed civilians in Spain against French occupation and regarded it as a model that his own country should have followed. He accepted the possibility that such a conflict would cause both sides to resort to extreme forms of violence against both combatants and noncombatants. For Clausewitz, guerrilla war constitutes an important element of his concept of the greater strength of the defensive, because it can lay the foundation for effective counterattack by a state whose regular army has been defeated even when much or all of the national territory has been occupied. Jomini, like Clausewitz, favors the organization of popular militias but rejects Clausewitz's advocacy of People's War. "The spectacle of a spontaneous uprising of a nation," he declares, "is rarely seen; and, though there be in it something grand and noble which commands our admiration, the consequences are so terrible that, for the sake of humanity, we ought to hope never to see it." Jomini's views were based on personal experience in Spain, where he had witnessed "reprisals, murder, pillage, and incendiarism throughout the country." He was thus convinced that "without being a utopian philanthropist, or a condottieri, a person may desire that wars of extermination may be banished from the code of nations, and that the defenses of nations by disciplined militia with the aid of good political alliances, may be sufficient to insure their independence" (ibid., 26, 30, 31).

Jomini's attempt to delegitimize guerrilla war was perhaps motivated by theoretical as well as humanitarian concerns. In *The Art of War,* he put forward what he called "The Fundamental Principle of War," which in a sentence amounts to the concept of winning great battles through the concentration of superior force. This principle, Jomini insisted, underlay "all operations of war" (ibid., 63). His prescription for success is not applicable, however, to guerrilla war. Clausewitz's recognition of guerrilla war as in the nature of things, and even as the means by which a defender can counter the effects of a great attacker success that has been produced by concentration of force, advances a serious challenge to Jomini's claim to have identified a universally applicable principle of war. The disagreement between Jomini and Clausewitz on the subject of guerrilla war is therefore about considerably more than what constitutes propriety in the conduct of armed hostilities.

Jomini tries to blunt the force of Clausewitz's forcefully argued contention that it is the "genius" of the commander-in-chief, rather than knowledge of principles and their application, that determines success in war. "The greater or lesser activity and boldness of the commanders of the armies are elements of success or failure," Jomini admits, "which cannot be submitted to rules." He then insists, however, that the great play of contingent factors in war does not invalidate the importance of principles, and moreover, that the significance of principles is demonstrated even when they are "applied accidentally" (ibid., 37, 38):

To relieve myself in advance of the blame which will be ascribed to me for attaching too much importance to the application of the few maxims laid down in my writings, I will repeat what I was the first to announce "*that war is not an exact science, but a drama full of passion*; that the moral qualities, the talents, the executive foresight and ability, the greatness of character, of the leaders, and the impulses, sympathies, and passions of the masses, have a great influence upon it."

Jomini, however, immediately vitiates, if not contradicts, his fervent protest by declaring, "I have not found a single case where these principles, correctly applied, did not lead to success" (ibid., 312–313; italics in original).

Preoccupied with the task of counterattack, Jomini portrayed Clausewitz as nothing more than a naysayer whose "incredulity" with respect to the existence of principles of war amounted to the "sapping of the basis of science" (*Art of War* [Winship/McLean], 15, 17). As a consequence, he did not engage *On War* on its own terms. But in the early years of the next century, an English writer with less to lose, and in need of ideas that corroborated his own predilections, used Clausewitz as the point of departure for his own exploration of the history and theory of war at sea.

Advocacy: Julian Corbett

Julian Stafford Corbett (1854–1922) was born in Great Britain, the second of six children of a wealthy architect and real-estate developer. He attended Trinity College, Cambridge University, where he received a first-class degree in law in 1875. Corbett was admitted to the bar in 1877, but his financial independence enabled him to interrupt his legal career with long trips to India and North America. Abandoning the law altogether in 1882, Corbett spent several years traveling before turning to writing fiction. Between 1886 and 1894 he published four novels, none of which found an audience. In 1889 and 1890, however, Corbett published short biographies of George Monk and Francis Drake for the Macmillan series "English Men of Action." Encouraged by the commercial success of these works, and recognizing that Drake was worthy of more serious treatment, Corbett devoted the next several years to research, reflection, and writing. The result was *Drake and the Tudor Navy,* published in 1898. This two-volume monograph established Corbett's reputation as a major naval historian. It was followed in 1900 by a sequel, *The Successors of Drake,* which was also greeted with acclaim.

Impressed by Corbett's work, Rear Admiral H. J. May invited him in 1902 to lecture in the War Course of the Royal Naval College, which then met at Greenwich (and was moved to Portsmouth in 1906 following a brief stay at Devonport). Corbett was to recall many years later that May instructed that his "leading line" should be "the deflection of strategy by politics" (Schurman, *Corbett,* 33).[14] Corbett's memory is probably accurate. "Expediency and strategy," May did indeed write to Corbett in August 1902, "are not always in accord" (ibid.). May's pithy statements resemble Clausewitz's idea that war is "an act of policy" rather than "a complete, untrammeled, absolute manifestation of violence" (*On War,* I/1: 87). It is possible that May had read Colonel J. J. Graham's English translation of *On War,* which had been published in 1873.[15] Corbett may have been primed to accept May's charge from the same source, or perhaps he came to the Prussian author's work while preparing his War Course lectures. Although the provenance of this line of thought is not known, however, there can be little doubt of its effect. Corbett's War Course lectures served as the basis for his next book, *England in the Mediterranean,* published in 1904. In this two-volume study, Corbett places much greater emphasis on the relationship between diplomacy and naval and military power than he had previously done. Clausewitz is not referred to by name, but in the penultimate paragraph in the preface Corbett makes an argument that is similar to his predecessor's contention in *On War* that outcomes could be determined by "engagements that did not take place but had merely been offered" (ibid., VI/8: 386).[16] After stating that comprehension of fleet action is impossible without explanation of diplomatic and military context, Corbett observes that "more often than not, the important fact is that no battle took place, and again and again the effort to prevent a collision is the controlling feature of widespread political action. As a rule, what did not happen is at least as important as what did."[17]

In 1906, Corbett added a lecture on "The System of Clausewitz" to his War College program, prompted, perhaps, by Captain Edmond John Warre Slade, the commandant of the War Course College, who was fluent in German and a serious student of German history.[18] Clausewitz's ideas are strongly evident in Corbett's "Strategical Terms and Definitions Used in Lectures on Naval History" of 1906, a booklet issued to students known as the "Green Pamphlet" after the color of its cover. In the Green Pamphlet, Corbett refers to Clausewitz by name. In addition, he argues, in agreement with Clausewitz, that the defense is superior in strength to the offense. "The Defensive, being negative in its aim," he writes, "is naturally the stronger form of war." A "true defensive," Corbett adds later, "means waiting for a chance to strike." Thus, he declares, "the strength and the essence of the defensive is the counterstroke." Corbett, like Clausewitz before him, identifies politics as an inhibitor of strategic action. The "deflection of strategy by politics," Corbett argues, though "usually regarded as a disease," is "really a vital factor in every strategical problem."[19] In the 1909 revision of the Green Pamphlet, these passages remain largely unchanged.[20]

In 1907, Corbett published his *England in the Seven Years' War,* a two-volume monograph that to an even greater degree than *England in the Mediterranean* focuses upon the interplay of diplomacy and naval and military power. Corbett justifies his approach with a direct reference to Clausewitz. In "modern times," Corbett observes, people habitually thought that war and peace represented two distinct conditions that meant "there was always a point where intercourse or diplomacy ended, and severance or strategy began." "Now Clausewitz," he argues, "with all the experiences of the Revolutionary and Napoleonic wars to guide him, long ago pointed out that this conception of international relations was false both in theory and practice." Corbett then declares that the English statesmen who are the subject of his study drew "no hard and fast line between diplomacy and strategy," and that indeed "every turn of hostilities presented itself diplomatically, and every diplomatic move as an aspect of strategy."[21]

Reference to Clausewitz's most famous aphorism is not, however, the only mark of the Prussian author's influence on Corbett. Although Corbett does not explicitly cite Clausewitz again in this work, to a remarkable degree it is clear that he chose his subject and cast his approach along lines that embodied some of the most important arguments that Clausewitz had put forward in *On War.* Corbett's style of writing about war changed, and it is not difficult to see that this is because he had begun to emulate Clausewitz.

Clausewitz, for example, favors rigorous consideration of a single case in order to provide readers with enough circumstantial detail to recover the difficult emotional conditions of strategic and operational decision-making surrounding an event. His primary concern is not right conduct or the correct application of rules or principles as such, but the character of command required to contend with the manifold difficult dilemmas of high command generated over the course of a conflict. Clausewitz recognizes the existence of principles of war, such as concentration of force, but he uses them as points of reference rather than standards of measure. In other words, the proper use of principles is to facilitate the understanding of the character of particular situations, not to serve as general instructions for action (*On War,* II/5–6).

Chapter One

Corbett's previous books had related the history of English naval operations, focusing on the development of English sea power as an instrument of state power over the span of many years—a generation in the case of the volumes on the Tudor navy (including the *Successors of Drake*), a century in the case of *England in the Mediterranean*. In effect, this resulted in something that resembled the general analysis of multiple wars in the manner of Jomini. *England in the Seven Years' War*, in contrast, examines strategic and operational decision-making during one conflict in great detail. This is a very Clausewitzian technique. In his first sentence, Corbett states his intention to "present Pitt's War as it was seen and felt by the men who were concerned with its direction." In order to accomplish this, he warns that he had found it "unavoidable to introduce a certain amount of strategical exposition." But he makes clear that even an excellent grasp of the principles of war is no guarantee of success. The "deepest notes" of the understanding of war, Corbett observes, "can only be heard when we watch great men of action struggling, as in some old Greek tragedy, with the inexorable laws of strategy, or riding on them in mastery to the inevitable catastrophe." Corbett knows that the task of writing this kind of history would tax those who read about past war "for its romance, for its drama and its poetry," but declares that "those who have ears for the real music of a great historical theme will not resent the sober cadences, without which it cannot be developed."[22]

Corbett's treatment of the political origins of the difference between limited and unlimited war also conforms closely to Clausewitz's views. Clausewitz argues that limited political objectives produce correspondingly limited—that is to say weak—military action, while conversely, unlimited political objectives, as would be generated in a conflict of national survival, generate the strongest possible forms of military action.[23] Corbett relates this idea to the changing circumstances of the Seven Years' War. The strategic objectives of the Duke of Newcastle, Britain's prime minister, at the outset of the Seven Years' War were relatively unambitious—defense of the home territory and minor territorial gains in North America. By the second year of the war, William Pitt, Newcastle's successor, had shifted to an offensive strategy against major French colonial interests around the globe in order to counterbalance the prospect of what seemed like all but certain French victory as part of a continental coalition over a militarily much inferior Prussia. The defeat of Prussia, a British ally, would have isolated Britain militarily and diplomatically and thus rendered it vulnerable to future French expansion in Europe and further abroad. "If Newcastle's measures seem weak," Corbett thus observes, "it must be remembered his object was limited. If Pitt at the end

seemed heroic, we must not forget his object had become practically unlimited—
it was a question of life and death between two empires, and the continued existence of France as a maritime power" (*England in the Seven Years' War,* I: 28–29).

The abject failure of Britain's defensive strategy at the beginning of the
Seven Years' War, and the stunning achievements of the offensive strategy that
succeeded it, could easily have served as the basis for the contention that the offense is a stronger form of war than the defense. Corbett would have none of it.
"Sound strategy for the King and Newcastle" during the skirmishing in North
America that preceded the outbreak of the war, he insisted, was defensive, given
the strength of the navy and the weakness of the army. This means "not mere
passive defense, but the true defensive strategy, that is, waiting till the pressure
of the conditions should present an opportunity for assuming the offensive with
effect" (ibid., I: 24–25). And by the same token, the fact of Britain's offensive
achievement is tempered by what Corbett regards as the near success of France's
defensive strategy. "The essence of the defensive," he writes in his conclusion, "is
waiting for an opportunity to pass to the offensive, and we cannot look back
upon the struggle which the French attitude so skillfully prolonged without a
shudder to see how nearly they were rewarded." "There is no clearer lesson in
history," Corbett then asserts, "how unwise and short-sighted it is to despise and
ridicule a naval defensive" (ibid., II: 373).

Corbett's next monograph, *The Campaign of Trafalgar,* was published in
1909. There are no direct references to Clausewitz, but Corbett's methodology
and analysis are nonetheless distinctly Clausewitzian. In the first place, the restricted chronology and narrow operational focus enables him to examine the
decision-making of Britain's naval high command in detail. Second, Corbett's
analysis centers on the intersections of diplomacy and military and naval action.
Third, he again discusses British behavior in terms of the containment of enemy
offensive thrusts followed by counterattack (Corbett, *Campaign of Trafalgar,* II:
256–257).[24] And finally, Corbett highlights the circumstances that produced the
dilemma, in the process justifying decisions that violated the fundamental principle of concentration of force. His anti-Jominian assessment is striking. In the
summer of 1805, the British government divided the British fleet to cover major
operations in Italy and South Africa, as well as providing for the security of
home waters. Although Corbett recognizes the enormous dangers posed by
such conduct, he concludes that "the whole question will serve as a warning that
the broad combined problems of Imperial defence are not to be solved off-hand
by the facile application of maxims" (ibid., II: 264).[25]

Chapter One

Corbett's most famous book, *Some Principles of Maritime Strategy,* was published in 1911. In this work, he provides a systematic presentation of his fundamental concepts of strategy. Thus ideas that had previously served as support for historical narrative became the main matter, with historical material reduced to the role of illustrative example. Corbett's purpose, moreover, is self-consciously polemical in the sense that he crafts his exposition to address certain critical strategic problems of his own time and place. His major arguments in favor of defensive action in home waters in order to allow offensive action in distant seas—operations that required cooperation between the army and the navy—and his concomitant downplaying of the importance of seeking out and destroying the main fleet of the enemy, in effect supported the radical strategic policy of Admiral Sir John Fisher, the service chief of the Royal Navy from 1904 to 1910.[26] Corbett, indeed, may have written his book at Fisher's request.[27]

In *Some Principles of Maritime Strategy,* Corbett makes Clausewitz's writing both an authoritative source and a foil for his own arguments. Corbett claims Clausewitz's backing when he argues that the proper role of theory is to "assist a capable man to acquire a broad outlook whereby he may with greater rapidity and certainty seize all the factors of a sudden situation" (*Some Principles*, 4). He quotes Clausewitz to support the proposition that theory creates "a common vehicle of expression and a common plane of thought" that makes efficient deliberations of a military staff or conference of the national leadership possible (ibid., 8). Corbett identifies the concept that war is "a continuation of policy by other means" as the prevailing fundamental theory of war. After naming Clausewitz as its inventor, he proceeds to explore the implications of the phrase, in so doing endorsing its validity and insisting upon its great significance (ibid., 17–19, 23–30). Corbett, as before, defines a proper defense in Clausewitzian terms, namely, as resisting enemy moves followed by counterattack, and he marshals a plethora of strong arguments in favor of the defense over the offense (ibid., 31–40).

Corbett does qualify his position, noting that he is "not in whole-hearted agreement with Clausewitz's doctrine of the strength of defence" (ibid., 72–73). He founds his own advocacy of maritime strategy as an effective form of limited war on Clausewitz's writing on the subject, but claims his views are different on a critical point. Corbett maintains that, unlike Clausewitz, he believes that the destruction of the enemy's main forces, under the conditions of limited war, need not be the primary objective. Clausewitz is in fact of the same opinion (*On War*, I/2: 92–93); Corbett, however, appears to have thought otherwise—he is convinced that Clausewitz's position is that action should always be directed at the enemy army. Clausewitz's supposed failure to recognize that his own concept of limited war could require action against objectives other than the main fighting forces of the enemy prompts Corbett to observe, "Clausewitz himself never apprehended the full significance of his brilliant theory" (*Some Principles,* 41–52).

In marked contrast to Clausewitz, and his own previous writing, Corbett sidesteps the issue of the ability of the commander-in-chief to direct war operations effectively:

> The conduct of war is so much a question of personality, of character, of common-sense, of rapid decision upon complex and ever-shifting factors, and those factors themselves are so varied, so intangible, so dependent upon unstable moral and physical conditions, that it seems incapable of being reduced to anything like true scientific analysis. (Ibid., 3)

Theory, he finally concludes, is a matter of deliberation and education, but "not of execution at all," which he notes depends upon "the combination of intangible human qualities which we call executive ability" (ibid., 6).

Corbett, for his part and following Jomini, states that the proper sphere of theory is the deliberations of the staff or national leadership (ibid., 6–8). Theory provides a common language—inculcated through education—that can facilitate productive discussion and enable decision-makers to recognize certain standards of right practice—what Corbett calls the determination of "the normal"—that can guide them to sound conclusions in particular cases. Corbett has reason to believe that this is Clausewitz's approach as well. In a note thought to have been written in 1830 that is included in *On War,* which Corbett quotes, Clausewitz indeed says that

> when it is a question not of taking action yourself, but of convincing others at the Council table, then everything depends on clear conceptions and the exposition of the inherent relations of things. So little progress has been made in this respect that most deliberations are merely verbal contentions which rest on no firm foundation, and end either in one retaining his own opinion, or in a compromise from considerations of mutual respect—a middle course of no actual value. (Ibid., 6, 9)[28]

Clausewitz preceded his remarks, however, with the observation that "all great commanders have acted on instinct, and the fact that their instinct was always sound is partly the measure of their innate greatness and genius. So far as action is concerned this will always be the case." And he followed the passage quoted by Corbett with the dismissive remark that "clear ideas on these matters do, therefore, have some practical value" (*On War*, 71).

In *Some Principles of Maritime Strategy,* in spite of Clausewitz's contrary position in Book II, Corbett characterized the executive ability of the commander-in-chief as a matter outside the purview of theory, and this may have reflected his actual views. But in a manner suggesting his sympathy with Clausewitz's treatment of the subject, in his historical monographs he had noted the human qualities that are required to resolve the problems posed by strategic dilemma. A similar apparent contradiction is to be found in *Some Principles of Maritime Strategy* when Corbett expresses an equivocal stance with respect to the proposition that defense is the stronger form of war, which marks a retreat from the uncompromising declaration of agreement with Clausewitz in the Green Pamphlet. In both cases, Corbett's later writing was probably influenced by political expediency.

Naval officers in the War Course resented what they regarded as Corbett's presumption in assuming the role of instructor of naval strategy, a subject in which he could claim no practical experience (Schurman, *Corbett,* 44–46, 50–59). This negative attitude would undoubtedly have been magnified considerably by any attempt on his part to speak with authority on the subject of executive capability in war, a matter that lay at the heart of their professional identity (ibid., 46). Corbett may have intentionally avoided the subject so that his arguments about strategic principles for a maritime power, which were of immediate importance to ongoing policy discussions, would be well received. But in following this course—if indeed this was the case—Corbett forfeited accurate representation of Clausewitz's major propositions about the psychological dimensions of high command and their implications for military theory. Naval officers believed as an article of faith that taking the initiative was preferable to waiting on events. They were therefore reflexively hostile to arguments that favored the defense over the offense. Corbett's more moderate statement of his views in this area was also probably intended to minimize opposition to his main propositions.

Corbett's representation of *On War,* while incomplete and occasionally faulty, nonetheless engages Clausewitz's thinking about the nature of strategic decision-making and about defense as the stronger form of war. His analysis is in many respects perceptive and his general evaluation highly favorable. Corbett's judgments, however, were to be overshadowed by the work of a younger countryman who held *On War* responsible for what was universally believed to be the greatest catastrophe in modern European history.

Repudiation: Basil Liddell Hart

Basil Liddell Hart (1895–1970) was born in France, the younger son of a Methodist minister. He entered Corpus Christi College, Cambridge University, in the fall of 1913 and studied history. His performance in the preliminary examination on his subject in May 1914 was undistinguished. Shortly after the outbreak of World War I in August 1914, Liddell Hart joined the British army, receiving a commission as second lieutenant. He served three short combat tours in France in 1915 and 1916. Liddell Hart was wounded by gas at the battle of the Somme in July 1916 during his third stint in the trenches. Debilitated, he served as an adjutant in training units for the remainder of the war. In 1918 he published a short book summarizing his experience as an instructor. Its success prompted him to produce a second book on tactics in 1921 and to begin writing for military journals. In 1923, Liddell Hart was released from active service on medical grounds stemming from the injuries he had suffered during the war. Within several years of leaving the army, he established a reputation as Britain's preeminent military journalist. Books on military history and theory soon followed.[29]

During the war, Liddell Hart had been an uncritical admirer of British military practice. His firsthand experience of fighting in the trenches and his thoughtful approach to training soldiers, however, prompted him to have second thoughts on the subject. The ideas that began to stir in him were much amplified by difficult encounters he had with senior officers after the war and by further reflection and study.[30] In *Paris, Or the Future of War*, published in 1925, Liddell Hart argues that Europe's prewar military leadership had mistakenly believed in two things: first, that victory in war requires the destruction of the main armed forces of the enemy; and second, that this was to be achieved through direct assaults with overwhelming numerical force, which required national conscription (Liddell Hart, *Paris or the Future of War*, 8–9).[31] These concepts, Liddell Hart asserts, are the essence of Napoleonic doctrine. He insists, however, that their disastrous influence on the military thinking of his own day had been brought about primarily through the work of Clausewitz.

Chapter One

It was Clausewitz, Liddell Hart explains, who had "analysed, codified, and deified the Napoleonic method." And it was Clausewitz, he further argues, who had attracted the devoted attention of the military students of Europe: "the master at whose feet" they had "sat for a century." And finally, it was from Clausewitz that the German army in particular had drawn "the inspiration by which they evolved their stupendous, if fundamentally unsound, structure of 'the nation in arms.'" German success against the French in 1870, he said, had caused the other great powers "to imitate the model, and to revive with ever greater intensity the Napoleonic tradition, until the gigantic edifice was put to an extended test in the years 1914–1918—with the result that in its fall it has brought low not only Germany, but, with it, the rest of Europe." Liddell Hart concedes that Clausewitz, though not his disciples, had recognized the existence of two objectives besides the enemy's military power, namely, its territory and will to resist. But he criticizes what he regards as the third-place priority given to the latter as a "vital mistake" (ibid., 10–11).

Liddell Hart characterizes the static trench stalemate on the Western front as the inevitable consequence of Clausewitzian thought. His antidote is maneuver warfare based upon the deployment of tanks, mobile infantry, artillery, and aircraft against the enemy's headquarters and communications. The objective of such a system of combined arms is not the physical destruction of the opposing army, but the dislocation and demoralization of its command. Blockade and strategic air attack, meanwhile, would strike at enemy morale at home. Liddell Hart believed that such a system of fighting would bring any war to a swift conclusion with much less expenditure of life and materiel than would be required in a war focusing on direct attack against the main forces. Military effectiveness, for this reason, was to be measured by the intelligent use of force to collapse enemy will in short order, not the sustenance of will to suffer losses greater than that of the enemy. In which case, Liddell Hart concludes, war based on "surprise and maneuver will reign again, restored to life and emerging from the mausoleums of mud built by Clausewitz and his successors" (ibid., 82–83).

Liddell Hart was apparently so pleased by his indictment of Clausewitz that he repeated these sentiments in virtually the same language two years later in *The Remaking of Modern Armies*.[32] In between, in the preface to his biography of Scipio Africanus, the great Roman commander, he observed that the "purblind apostles of Clausewitz" had "deceived themselves . . . and the world" that slaughter was "synonymous with victory."[33] In his next three books, which were devoted to historical biography, he avoided overt discussion of his theoretical ideas and did not mention Clausewitz.[34]

Liddell Hart's attacks on Clausewitz prompted Spenser Wilkinson, then a septuagenarian but still highly respected as a military analyst, to challenge what he regarded as the mischaracterization of the Prussian theorist's ideas in the fall of 1927.[35] Wilkinson's powerful critique may have caused Liddell Hart to restrain momentarily his attacks on Clausewitz. In Liddell Hart's book *The Decisive Wars of History: A Study in Strategy,* published in 1929, references to Clausewitz are few and short, and in general more temperate than in his earlier work.[36] Liddell Hart criticizes Clausewitz's definition of strategy on the grounds of being open to misinterpretation rather than being in essence unsound, although he does refer dismissively to "the dogma of Clausewitz that blood is the price of victory" (*Decisive Wars,* 147, 203). In *The Real War, 1914–1918,* a survey history of World War I published in 1930, Liddell Hart does not mention Clausewitz, which in light of what he had written previously—and what he would write in the future—is perhaps significant.[37]

Liddell Hart's restraint, if such it was, with regard to Clausewitz did not last. His study of Ferdinand Foch, an apostle of the offensive before World War I and commander-in-chief of the Allied armies at its conclusion, marks a return to vigorous denunciation. Foch had praised Clausewitz's *On War* for being "solid," by which he meant that he regarded it to be highly substantive. In *Foch: The Man of Orleans,* published in 1931, Liddell Hart observes that

> the ponderous tomes of Clausewitz are so solid as to cause mental
> indigestion to any student who swallows them without a long course
> of preparation. Only a mind developed by years of study and
> reflection can dissolve the solid lump into digestible particles. Critical
> power and a wide knowledge of history are also necessary for
> producing the juices to counteract the Clausewitzian fermentation.
> (*Foch,* 133)[38]

Foch, Liddell Hart argues, "had caught only Clausewitz's strident generalizations, and not his subtler undertones" (ibid., 23). He also castigated General Alfred von Schlieffen, famed designer of the German offensive strategy in 1914, for being a "fervid devotee of Clausewitz" and thus pressing "to its logical extreme the doctrine that military success cancelled all other factors" (ibid., 75). Liddell Hart criticizes General Erich von Ludendorff, the architect of the German spring offensive of 1918, for relying on direct assault on strong points rather than maneuver, thus showing that in spite of his declared intention to do otherwise, "he could not free himself from the dead hand of Clausewitz" (ibid., 282). Liddell Hart concludes that Foch's theory of war "was but the reflection of Clausewitz, save for certain contradictions original to himself," which in effect "made strategy merely the servant of tactics" (ibid., 473). In *The British Way in Warfare,* published in 1932, Liddell Hart again holds Clausewitz accountable in part for the excesses of World War I by linking him to Foch, condemns Clausewitz for "advocating the principle of unlimited violence," criticizes Clausewitz's "tendency to dramatic generalization which obscured his many discerning reflections" (*British Way,* 17–20),[39] and finds fault with Clausewitz's definition of strategy as being concerned with the use of battles on the grounds that it promotes the diminution of government responsibility for the conduct of war (ibid., 95).

In 1932, Liddell Hart delivered the Lees Knowles lectures at Trinity College, Cambridge. These were revised and published as *The Ghost of Napoleon* in 1934. In this work, Liddell Hart assigns Clausewitz the role of master villain in a morality play about the role of ideas and the conduct of war.[40] He begins with a general proposition, namely that "the influence of thought on thought is the most influential factor in history" (*Ghost of Napoleon,* 11). By this he means that military concepts can shape the policies of governments in ways that can have momentous effects, even forwarding the progress of science and the arts. Liddell Hart conceded that the stimulation of science and art by new methods of warfare that were the product of intellectual advances had received a great deal of scholarly attention. But he noted that "men of thought who produced ideas of a more concrete nature, whose thoughts more directly influenced the course of history, have been comparatively overlooked. And their influence on events has not been studied with a due sense of proportion" (ibid.).

He uses two main arguments to illustrate this thesis. First, Napoleon's great successes were attributable to his exploitation of advances in maneuver warfare as generated by the work of Marshal Saxe, Pierre de Bourcet, and Comte de Guibert (ibid., 27–100). And second, Clausewitz, by mischaracterizing the warfare of his generation in terms of the maximization of violence and the opponent's army as the primary objectives of strategy, obliterated the memory of the Napoleonic war of movement and thus laid the foundation for the static conditions of fighting on the Western Front during World War I. "Although Clausewitz used Napoleon freely as an example, and came to be taken by the world as the interpreter of Napoleon," Liddell Hart argues, he really expressed ideas that originated in his own mind. He was the prophet, not of Napoleon, but of himself. If one weighs his influence and his emphasis, one might describe him historically as the Mahdi of mass and mutual massacre. For he was the source of the doctrine of "absolute war", the fight to a finish theory which, beginning with the argument that "war is only a continuation of state policy by other means", ended by making policy the slave of strategy. (Ibid., 120)

Liddell Hart is incensed by what he believes to be Clausewitz's emphasis on the importance of maximizing violence to fight and destroy the enemy's main army. "Clausewitz's principle of force without limit and without calculation of cost," he declares, "fits, and is only fit for, a hate-maddened mob. It is the negation of statesmanship—and of intelligent strategy, which seeks to serve the ends of policy" (ibid., 122).[41] Liddell Hart is convinced that the use of the greatest possible force without regard to losses is of central importance to Clausewitz:

> The conception of "absolute war", which formed the keystone of his doctrine was the most extreme, and the most unreal, of all his contributions to thought. For if the term "absolute war" has any meaning it is that of fighting until the capacity of one side for further resistance is exhausted. In practice, this may well mean that its conqueror is brought to the verge of exhaustion, too weak to reap the harvest of his victory. (Ibid., 143)

"Clausewitz," Liddell Hart thus observes, "looked only to the end of a war, not beyond war to the subsequent peace" (ibid., 121).

As other studies have demonstrated,[42] Liddell Hart's assessments of Clausewitz's writing are unfair. Wilkinson's commentary, and even his own reading, must have made him aware that his treatment of the Prussian author placed him on dangerous ground. Liddell Hart thus explains, as he had before, that Clausewitz is guilty of writing in a manner that is highly susceptible to misinterpretation, rather than of faulty reasoning as such. "Clausewitz," he observes, "was so difficult to follow that his emphatic generalizations made more impression than his careful qualifications" (*Ghost of Napoleon*, 105). "Unfortunately," Liddell Hart later notes,

> his qualifications came on late pages, and were conveyed in a philosophical language that befogged the plain soldier, essentially concrete minded. Such readers grasped the obvious implication of the

leading phrases, and lost sight of what followed owing to distance and obscurity. (Ibid., 123)[43]

"Not one reader in a hundred," he thus concludes, "was likely to follow the subtlety of his logic, or to preserve a true balance amid such philosophical jugglery" (ibid., 125).[44] And yet, Liddell Hart makes clear in his late correspondence that he believes that Clausewitz's writing prescribes courses of action with reprehensible characteristics unmitigated by significant qualification (Gat, *History of Military Thought*, 692).

In print, Liddell Hart argues that Clausewitz is accountable for the consequences of the misreading of his writing, in spite of whatever nuanced meaning existed in the actual text:

> In justice to Clausewitz, one must draw attention to his reservations, but for true history one must concentrate attention on his abstract generalizations, because it was the effect of these that influenced the course of European history. Moreover, Clausewitz himself had a direct as well as an indirect responsibility. For while he saw the limitations which reality placed upon the abstract ideal, he tended to set up the latter as his ideal in the actual conduct of war. He seemed to think that, by pursuing the extreme, a commander would come nearest the practical mean. The result, however, was that in exalting logic he dethroned reason and encouraged his disciples to break away from reality. (*Ghost of Napoleon*, 123–124)

"Not merely stalemate," Liddell Hart expounds, "but massed suicide— more truly, homicide—was the penalty of Clausewitz's theory of mass" (ibid., 129). He reasons, therefore, that Clausewitz bears "a direct responsibility for the ruinous cost and negative nature of the War of 1914–1918" and "a significant, if less direct, responsibility for bringing about that war" (ibid., 13). "Never, surely," he later concludes, "has a theory had so fatal a fascination" (ibid., 144).

The Ghost of Napoleon is Liddell Hart's longest and strongest statement of his views on what he believes are Clausewitz's wrongheaded ideas and their iniquitous consequences.[45] Subsequent books do little more than repeat its criticisms in the course of addressing other matters.[46] Liddell Hart's most widely read book, *Strategy*, published in 1954, did not add much to his previous criticism, but its litany of complaint contained memorable phrases that almost certainly influenced the opinions of many readers. Clausewitz, he argues, "acquired a philosophical mode of expression without developing a truly philosophical mind." He "contributes no new or strikingly progressive ideas," Liddell Hart contends, "to tactics or strategy," and is in this sense "a *codifying* thinker rather than a *creative* or *dynamic* one." "In defining the military aim," he insists, "Clausewitz was carried away by his passion for pure logic." Had Clausewitz lived longer, Liddell Hart surmises, he would have dropped "his original concept of 'absolute' war" and revised "his whole theory on more common-sense lines" (*Strategy*, 352–357; italics in original).

Liddell Hart's characterization of Clausewitz as a blood-thirsty proponent of the attack at all costs is based in large part upon his misreading of the Prussian author's statements about the primary objective of war. In the first chapter of Book I, Clausewitz writes that "the destruction of the enemy's forces is always the means by which the purpose of the engagement is achieved," but he also writes that "the purpose in question may be the destruction of the enemy's forces, but not necessarily so; it may be quite different" (*On War*, I/1: 95). In the second chapter, he does maintain that the opposing "fighting forces must be *destroyed: that is, they must be put in such a condition that they can no longer carry on the fight*" (ibid., I/1: 90; italics in original). This declaration, however, is made with respect to a hypothetical form of absolute war in which politics is not a governing factor. War in reality could approach its absolute form for a time, but Clausewitz is convinced that all real wars are ultimately about politics. Thus, he argues, the "aim of *disarming the enemy . . .* is in fact not always encountered in reality, and need not be fully achieved as a condition of peace. On no account should theory raise it to the level of a law" (ibid., I/2: 91; italics in original).[47] Clausewitz later explains:

> We can now see that in war many roads lead to success, and that they do not all involve the opponent's outright defeat. They range from *the destruction of the enemy's forces,* [*to*] *the conquest of his territory, to a temporary occupation or invasion, to projects with an immediate political purpose, and finally to passively awaiting the enemy's attacks.* Any one of these may be used to overcome the enemy's will: the choice depends upon circumstances. (Ibid., I/2: 94; italics in original)

Much of Liddell Hart's confusion is undoubtedly attributable to his understanding of Clausewitz's characterization of battle as the "center of gravity" of armed conflict without due attention to the Prussian author's qualification of his argument. In Book IV, Clausewitz declares that "battle must always be considered as the true center of gravity of the war. All in all, therefore, its distinguishing feature is that, more than any other type of action, battle exists for its own sake alone." He subsequently limits the force of this statement, however, as follows:

> If a battle is primarily an end in itself, the elements of its decision must be contained in it. In other words, victory must be pursued so long as it lies within the realm of the possible; battle must never be abandoned because of particular circumstances, but only when the strength available has quite clearly become inadequate. (Ibid., IV/9: 248)

In Book VIII, Clausewitz defines "center of gravity" as "the hub of all power and movement" and thus "the point against which all our energies should be directed." This could be the enemy army, his capital, or the enemy's principal ally, "if that ally was more powerful than he." Although Clausewitz recommends "the defeat and destruction" of the enemy through battle as the best course, it is not the only course, and certainly not to be entertained if the forces available are inadequate (ibid., VIII/8: 595–597).

Clausewitz's statements about the basic importance of battle are a definitional point of reference, not a recommendation for a particular course of action. The purpose of his remarks is to make clear his opposition to the notion that war could be understood without connecting it to either the actuality or the prospect of serious fighting (ibid., IV/3: 228). Clausewitz is convinced that the *idea* of serious fighting is fundamental to war—that is, all action in war has to be taken on the assumption that a major battle could happen, even if it does not in fact occur. He thus writes:

> Combat is the only effective force in war; its aim is to destroy the enemy's forces as a means to a further end. That holds good even if no actual fighting occurs, because the outcome rests on the assumption that if it came to fighting, the enemy would be destroyed. It follows that the destruction of the enemy's force underlies all military actions; all plans are ultimately based on it, resting on it like an arch on its abutment. Consequently, all action is undertaken in the belief that if the ultimate test of arms should actually occur, the outcome would be *favorable*. The decision by arms is for all major and minor operations in war what cash payment is in commerce. Regardless how complex the relationship between the two parties, regardless how rarely settlements actually occur, they can never be entirely absent. (Ibid., I/2: 97; italics in original)[48]

Clausewitz later makes it clear that he is not in favor of seeking battle as an end in itself. "Blind aggressiveness," he explains, "would destroy the attack itself, not the defense, and this is not what we are talking about." And he warns that the danger of trying to destroy the enemy's forces "is that the greater the success we seek, the greater will be the damage if we fail" (ibid., I/2: 97).

Liddell Hart does not fault everything that Clausewitz wrote (or that he thought he wrote). He approves of Clausewitz's views on the importance of considering the psychological circumstances of decision-making for a commander-in-chief under difficult conditions, the implications of this for the writing of military history, and the strategic superiority of the defense over the offense. In *The British Way in Warfare,* summarizing Clausewitz's arguments on the importance of command psychology and its implications,[49] Liddell Hart argues that military history

> should be directed mainly to discover the commander's thoughts and impressions and the decisions which sprang from them. To explore all the details of the fighting is unnecessary, valueless and even misleading. For it matters little what the situation actually was at any particular point or moment; all that matters is what the commander thought it was. (*British Way,* 108)

"Military history," Liddell Hart writes in his conclusion to this line of thought, "to be of practical value should be a study of the psychological reactions of the commander, with merely a background of events to throw their thoughts, impressions and decisions into clear relief" (ibid.).[50]

In *The Ghost of Napoleon,* Liddell Hart expresses his admiration of Clausewitz's emphasis on this theme, declaring:

> Clausewitz's greatest contribution to the theory of war was his elucidation of the moral sphere. Raising his voice against the mathematical school, he showed that the human spirit was infinitely more important than lines or angles. He discussed the effect of danger and fatigue, the value of boldness and determination, and with deep understanding.[51]

Liddell Hart also recognizes that Clausewitz argued that defense is the stronger form of war. In *The Defence of Britain,* published in 1939, he notes that Clausewitz believed that "the defensive was the stronger form of action."[52] He maintains, moreover, that the most effective strategy is defense followed by counterattack, which was also Clausewitz's view, although Liddell Hart does not seem to be aware that this was the case. In *The Decisive Wars of History*, he argues in favor of "elastic defence—calculated retirement—capped by a tactical offensive," a proposition that he believes was a manifestation, "in a deeper and wider sense than Clausewitz implied, that the defensive is the strongest as well as the most economical form of strategy."[53] Liddell Hart later states explicitly—and incorrectly—that Clausewitz did not recognize that counterattack could convert a successful defense into a positive act. Clausewitz, he writes in *Thoughts on War*, in a section composed in October 1937,

> took care to point out that the defensive was the stronger form of
> action, although he could not see how, of itself, it could produce a
> decision. That was true so far as the positive overthrow of the enemy
> was necessary for the fulfillment of the purpose in war. Even so there
> has always been a method by which the immediate advantages of the
> defence and the ultimate advantages of the attack could be combined.
> History offers, to those who will inquire of it objectively,
> overwhelming evidence that the counter-offensive, after the enemy
> has overstrained himself in the offensive, has been the most decisive
> form of action.[54]

As should be clear from the above discussion, Liddell Hart's descriptions of Clausewitz's views on the defensive are in certain important respects inaccurate. In addition, his comments on this matter as well as on the psychology of command-decision and the writing of history are noted in passing rather than recognized as the fundamentally important subjects that they are. There are several reasons for these flaws in his analysis. In the first place, Liddell Hart seems to have given *On War* a less than rigorous read in his early years of writing, and he never corrected for this through careful rereading later on.[55] In the second place, Liddell Hart believed that Clausewitz epitomized a school of military thought that had produced catastrophic effects in the immediate past, and that the continued authority of Clausewitz's work threatened more of the same in the future; in order to extirpate his influence, Liddell Hart sacrifices critical perspective.[56] And in the third place, Liddell Hart's agreement with Clausewitz's views on the importance of moral factors and the superiority of the defense over the offense is counterbalanced by his disagreement with him over how such concepts should be applied. This third point requires some explanation.

Insofar as the moral factor is concerned, Liddell Hart, in both *The British War of Warfare* and *Thoughts on War*, argues that "the teaching of Clausewitz was directed more to fortify the will of the commander on his own side than to undermine the will of the opposing commander."[57] Clausewitz indeed identifies the will of the commander-in-chief as a critical variable, and he focuses his analysis upon the general circumstances in war that challenge a general's psyche (*On War*, I/3: 7; VII/22: 573). Liddell Hart instead wants to explain the existence of specific means of dislocating the moral equilibrium of an opposing commander to decisive effect. As for the superiority of the defense over the offense, Clausewitz believes that the fundamental strength of the defense lay in the ability of the defender, given certain conditions, to carry out a prolonged resistance by mobilizing the manpower of the nation (ibid., I/1: 79–80; III/3: 192; VIII/3: 592–593). In the event of the defeat of the defender's organized forces and occupation of most or all of its territory, the defender could resort to guerrilla warfare (ibid., VI/26). Liddell Hart, in contrast, abhors protracted conflict,[58] opposes national conscription, and in particular has serious misgivings about guerrilla warfare.[59] His concept of defense calls for swift decision based upon the military virtuosity of a small professional army.

In *The Defence of the West,* published in 1950, Liddell Hart notes that although Clausewitz's writings are the "'Holy Scriptures' of the military profession . . . few of its members ever study them thoroughly."[60] It is possible that he saw himself as guilty of this charge. Here honest introspection, which was one of Liddell Hart's characteristics, may have prompted what Alex Danchev has called his capacity for "cryptic confession."[61] In any case, the popularity of Liddell Hart's engaging prose and vigorous argument meant that his views on Clausewitz were disseminated widely and to considerable effect.[62] In 1976, however, serious scholarship provided large correctives for much that is misleading or incorrect in his work.

2

Scholars

FOR CLAUSEWITZIAN STUDIES, 1976 was a seminal year. Michael Howard and Peter Paret published a new English translation of *On War,* which was more accurate and much more readable than its predecessors.[1] Paret also published *Clausewitz and the State,* a study of the political and intellectual background of the Prussian author's writing based upon a thorough examination of the relevant contemporary literature.[2] Raymond Aron published *Thinking about War: Clausewitz* (henceforth referred to by its English-language title, *Clausewitz: Philosopher of War*), a comprehensive and rigorous consideration of the meaning and implications of Clausewitzian political and military theory.[3] And finally, W. B. Gallie delivered the Wiles Lectures at Queen's University, Belfast—later published as *Philosophers of Peace and War*—which included a talk on *On War* that provided a penetrating philosophical assessment of its critically important opening chapter.[4]

The rich harvest of 1976 did not result in a consensus view of the general meaning of *On War*; indeed, interpretations of the work have varied widely and its meaning remains open to debate. This scholarly work did, however, accomplish four tasks that were prerequisites to further major progress. First, the Howard/Paret edition of *On War* established a complete translation that is accurate enough to be generally accepted as a sound starting point for inquiry and discussion by English-language readers. Second, Paret provided a thorough investigation of the *literary* sources of Clausewitz's thought and its development. Third, Aron's examination of the text of *On War* set appropriate terms of textual analysis, posed many of the right questions about theoretical substance, corrected a number of previous errors, identified important problem areas, and offered a host of perceptive insights. And fourth, Gallie's intelligent consideration of Clausewitzian philosophical method and his highly suggestive critical reasoning pointed to promising new lines of inquiry.

Although the work of Aron, Paret, and Gallie marked major advances in Clausewitzian studies, none of these thinkers achieved complete command of

On War. Nor has anyone since their time produced a satisfactory explanation of Clausewitz's thought. To understand why, we must undertake a detailed examination of the strengths and weaknesses of what was accomplished in 1976. This task will begin with Aron, whose study of Clausewitz set the standard for direct assault on the enigma known as *On War*.

Text: Raymond Aron

Raymond Claude Ferdinand Aron (1905–1983) was born in France, the last of three sons of a middle-class Jewish family. A gifted scholar, he graduated at the top of his class at the Sorbonne in 1928 with a thesis on Kant. He served as an enlisted man in the French army from October 1928 to March 1930, and from the spring of 1930 to mid 1933 he taught at the University of Cologne. There, he developed fluency in German and a command of German philosophy and social science. Over the course of 1933 and 1934, while teaching in Le Havre and Paris, he completed his first book, a study of German sociology. By the spring of 1937, Aron had finished two additional books on the philosophy of history. In 1938 he received his doctorate from the Sorbonne. War interrupted his teaching career and he served briefly as a technician in the army. Following the German victory in the spring of 1940, Aron left France for Great Britain. His success as a political writer in the service of the Free French cause from 1940 through 1945 led him to abandon academia to become a political journalist. From 1946 to 1955, Aron became famous as one of France's leading commentators on current affairs. In 1955, he became a professor at the Sorbonne.[5]

Aron first encountered the work of Clausewitz through discussions with Herbert Rosinski (1903–1962) in Berlin in 1931.[6] Rosinski was convinced that *On War* could only be understood by taking account of the character of Clausewitz's intellectual development, and in particular, the direction of his thought in his last years. During World War II, Aron's interest in Clausewitz was stimulated in London by Stanislas Szymonczyk, an exile like himself who was also a Free French journalist. Szymonczyk, Aron recalls in *Clausewitz: Philosopher of War*, "readily made use of Clausewitzian phrases to enhance the tone and style of strict analysis" (*Clausewitz*, vii). Thus, he confesses, *On War* "was for me, as for so many others, a treasury of quotations" (ibid., viii). Aron did not read *On War* until its appearance in a French translation in 1955. At first, he attempted to use Clausewitz's statement that war was an act to compel an opponent to comply with one's will as the basis of his own theory of war. Aron found this approach unsatisfactory, however, and decided that critical comprehension of *On War* required careful and prolonged study. In academic year 1971–1972, Aron taught a course on Clausewitz at the Sorbonne in which he began his serious explorations. From 1972 to 1975, he states in his memoirs, he worked "with alacrity, almost with enthusiasm" (*Memoirs,* 408–409).

Aron admits that he found Clausewitz to have an intellectual and political spirit compatible with, and in important respects similar to, his own. "As nobody," he maintains, "who spends years in a kind of intimate dialogue with another mind can help seeing things in the light of that silent and insistent interlocutor, it is only right that I should confess my sympathy with him" (*Clausewitz,* ix). But Aron's feelings of connection with Clausewitz were based on more than scholarly exchange. His years of teaching in Germany between the world wars had produced a deep knowledge of and disposition favorable to German philosophy and sociology (*Memoirs,* 42–50). The success of his marriage in the shadow of French political disorder and decline in the late 1930s had prompted Aron to harbor feelings of "private happiness, public despair" (ibid., 53)—a state of mind that mirrored Clausewitz's for much of his adult life. Like Clausewitz, Aron had suffered exile and political estrangement from his own country's government following military disaster. "A Frenchman has only to remember his own experiences between 1940 and 1945," Aron writes, "to sympathize with the attitude of Clausewitz between 1806 and 1815" (*Clausewitz,* x). And perhaps above all, Aron, like Clausewitz, had devoted his life after the great wars of his time to reflecting on their nature and practical political and military implications (ibid., ix).

Aron's approach to Clausewitz is fairly straightforward. "To interpret him," he declares, "is above all to understand what he said, starting from the sensible hypothesis that he said what he wanted to say." Aron notes, however, that *On War* is made up of sections that were written at different times, and that examination of concepts therefore had to take into account "an ordering of texts according to their date." Moreover, one had to "approach the whole" and not "take phrases out of their contexts." Aron warns that readings of the text must also address Clausewitz's particular concerns about his own era. "This perfectly historical means of interpretation," he explains, "does not exhaust the content of any great work, but it nevertheless averts what continues to appear to me to be an intellectual error and an ethical fault: the translation of the thought of an author into a language or conceptual system which was either foreign to him or assumed a different tone (or a different meaning) in his age than it does in ours" (ibid., 2–3).

Aron wonders what Clausewitz intended "to say to posterity." Indeed, his desire to use Clausewitz's thoughts to address critical problems of his own day was his strongest motive for trying to come to terms with *On War*. But it is essential, he insists, first to determine "what Clausewitz meant to say to those who belonged to his world, to those who shared the same historical experience and who gave the same meanings to words." Only then could one discover "the interpretation of the meaning or meanings which his work and system retain or take on for us, in terms of our own world and our own experiences." Aron argues that to apply Clausewitz's words in support of contemporary arguments in ways that would violate their original sense would amount to little more than making language signify whatever suits the needs of the moment. Such action, he is convinced, has little intellectual merit. "What is the good of questioning an interlocutor," he observes, "who will only send us back our own words, like an echo?" As for himself, Aron declares that his examination of Clausewitz is based upon "my willingness to read and to listen while detaching myself from my own position—perhaps an unattainable ideal but not an inconsistent desire" (ibid., 3, 6).

Aron believes that "the purpose of Clausewitz can be easily seen if you are prepared to read him carefully," but that "there have not been many careful readers." At a general level, he argues that Clausewitz "sought to formulate a conceptual system, a theory (in the sense that we speak of economic theory today) which enables the concept of war (or real wars) to be thought out with lucidity." More specifically, Aron recognizes that Clausewitz claimed he had invented a nonprescriptive theory of war that was of great practical significance. It was indeed the elucidation of exactly what the nature of this formulation was that prompted Aron's serious inquiry into the meaning of *On War* in the first place. "What attracted me initially," Aron writes, "was the philosophical problem, the effort required to grasp the nature of war, to formulate a theory which would not be confused with a doctrine, in other words which would teach the strategist to understand his task without entertaining any absurd claim to communicate the secret of victory" (ibid., viii, 5).

Aron does not believe that consideration of the work of Kant, Hegel, and Montesquieu—whose writing he knew well (*Memoirs*, 25–27, 68–69, 240–241)—could serve as the basis for understanding Clausewitz. He admits that there is evidence leading to "a prudent, banal conclusion that Clausewitz probably knew Kantian thinking, at least indirectly," but he rejects the argument that Clausewitz was either a disciple of Kant or applied Kant's ideas. Aron is no less skeptical of the claim that *On War* embodies Hegel's dialectical manner of reasoning. "The essence of Hegel's historical dialectic is the synthesis that rises above temporal contradictions and gives a rational significance to the progression," he argues, "but it has no place" in *On War*. "It might be suggested that Clausewitz's conceptual dialectic, as distinct from the historical one," Aron adds, "is close to Hegel's, but this is clearly not the case." He then goes on to say that "it might be possible to describe as dialectical the distinction between the polarity concerning the two sides' desire for victory and the non-polar relationship of attack and defence, but it hardly stems from any specific part of the Hegelian dialectic" (*Clausewitz,* 226–230). Aron does maintain that "the conceptualization of Clausewitz resembles that of Montesquieu far more closely than anybody has ever suggested, and far more than any similarity it may bear to the works of Hegel or Kant" (ibid., 54).[7] But, he concludes, "I do not think that the influence of Montesquieu matters very much" (ibid., 232).

The key to Clausewitz's thinking, Aron is thus convinced, must be found in the originality of his thought, as evidenced in his writing, rather than in the identification of imported methodology. This called for the careful study of not only *On War* but much of Clausewitz's other work, which is considerable, and a large body of critical literature as well. Aron writes,

> I spent time in the Bibliotheque Nationale, I read the French and
> German commentators on the treatise; I took pleasure in
> rediscovering old quarrels. I did not claim that I was putting an end to
> the controversies, but I was reconstructing Clausewitz's intellectual
> path through the various versions of his great work. (*Memoirs,* 409)

Aron's completed book reflects this approach. It consists of two volumes (published together in the English translation). The first deals with what Clausewitz tried to accomplish in *On War,* the second with the significance of his thinking in the twentieth century. The subject of the second volume does not concern the present inquiry and will not be considered here, except in reference to Aron's assessment of Liddell Hart. Of the first volume, Aron writes, "On the whole I kept strictly to the texts themselves and to the military writers known by Clausewitz, and I willingly disregarded the commentators except to set their interpretations against mine" (*Clausewitz,* 9).

The first volume of *Clausewitz: Philosopher of War* is divided into three parts. Part I, entitled "From Writer to Writings," consists of a biographical sketch, a description of Clausewitz's intellectual development, and Aron's conclusions about Clausewitz's views on war as they stood at the end of his life. Clausewitz, Aron argues, was from the very first hostile to the notion of prescriptive theory:

> At the age of 25, influenced by Scharnhorst and events, he already knew which types of theory to reject as being contrary to the nature of things and as offering bad advice: namely those which failed to recognize the role of emotion, of military virtues and of passions; in short, of the human side of war and its conduct, those which put forward strict rules and claim to have discovered one rule amongst them all which is responsible for victory or defeat, those which failed to take heed of the singularity of each combination of events, and exclude the part played by accident and good or bad luck. (Ibid., viii)

According to Aron, Clausewitz avoids the formulation of prescriptive doctrine in his earlier work by describing all the factors governing the prosecution of war without favor. Such an approach, he argues, "permits the definitive refutation of dogmatism since all the elements of strategy find their place in the whole, and consequently none of them can claim an exclusive or disproportionate importance" (ibid., 51).

"In all texts written before 1827," Aron writes, "political considerations are not absent but they never seem to influence the conduct of war directly." But Clausewitz in his last years came to two important realizations: first, that politics heavily influences operations during the course of a war, and second, that, this being the case, the question of the ends of war—namely, "to overthrow the enemy or not"—is at all times not a strictly military question but a political one as well. This meant that "suddenly the whole of what he had thought and written for more than twenty years required revision"; the characterization of ends as being contingent unavoidably required consideration of the means—that is, the degree of fighting—as contingent as well. "This is the real problem which has puzzled the commentators," Aron declares, "and which, in my opinion, obsesses Clausewitz in the course of his last years, between 1827 and 1830" (ibid., 59).

The "real" problem Aron refers to is the question of what determines whether a war is limited or unlimited. In other words, does one "impose peace or negotiate it," pursue "victory by 'knockout' or victory on points"? In reality, Aron observes, Clausewitz's formulation allows for the existence of "many intermediaries" in between these outer limits (ibid., 60, 83). But this does not resolve what appears to be a fundamental contradiction in Clausewitzian theory as it is articulated in the first chapter of Book I of *On War,* namely, the assertion, on the one hand, that in theory the essential nature of war is the maximization of violence, and the recognition, on the other, that war in practice is an instrument of policy. The former seems inescapably to favor unlimited war, whereas the latter, depending on circumstances, can countenance either.

The primary interpretive problem for Aron is defined by his belief that Clausewitz completed the first chapter of Book I after writing all the other sections of *On War.* Thus, he presumes that Clausewitz's new ideas about the influence of policy on war did not inform the bulk of the treatise. If he is correct about this, it means that the critical antinomy cannot be resolved directly through consideration of the text following the opening chapter, because it is not addressed there by Clausewitz in the first place. Aron's solution to this conundrum is to use the supposedly new ideas that are expressed in the first chapter of Book I to modulate his understanding of concepts described in the text that follows, for which he could claim the authority of Clausewitz's own instructions of 10 July 1827 and the unfinished note supposedly written in 1830 for the incomplete draft of *On War.* "Since, as I believe," Aron explains, "Clausewitz only mastered his own system at the time when he wrote the first chapter of Book I, it is advisable to read and interpret the whole in the light of what I sometimes call the final synthesis or intellectual testament" (ibid., 9).

This approach leads Aron to engage what he regarded as the salient characteristics of Clausewitz's thought, with Book I, Chapter 1, given prominence of place, rather than to undertake a formal exegesis of *On War* as a coherent whole. Part II of Aron's first volume, entitled "The Dialectic," examines the character of Clausewitzian reasoning through consideration of three dialectical conceptual pairs. The first of these considers the instrument of war, on the one hand, which is fighting, and the end of war, on the other, which is policy (means-end); the second takes into account the moral and physical factors that govern the prosecution of a war (moral-physical); and the third, the nature and relative merits of defense and attack (defense-attack). Part III, entitled "The Theoretical Scheme," evaluates the text as theory. It thus considers the function of Clausewitz's propositions within what is supposed to be a nondoctrinal system of theory; the degree to which Clausewitzian theory is susceptible to degradation because of changes in circumstances over time; and the practical character of Clausewitzian theory.

In Parts II and III, Aron subjects his chosen issues to close scrutiny and searching analysis. The intelligence and vigor of the inquiry notwithstanding, however, he fails to produce a convincing synthetic explanation of Clausewitzian thought. Aron's findings are unsatisfactory in three critical areas. First, he is unable to resolve what he believes is the apparent contradiction between two of Clausewitz's contentions: that war as an ideal calls for the use of the maximum amount of violence, and that war in reality does not usually take this form. To the question, "Why does real war not conform to its own nature?" Aron replies simply that "the Treatise remains unfinished and does not resolve its own enigma" (ibid., 83).[8] Second, although Aron recognizes Clausewitz's claim to have invented an antidogmatic form of military theory that is of great practical value, his description of what it is amounts to little more than platitudes. And third, although Aron acknowledges that the strength of the defense with respect to the attack is a key component of *On War,* his explanation of why this is so is equivocal, leaving major questions unanswered. I shall address the first problem, which is partially solved by Paret and Gallie, in the sections on these writers later in this chapter. The second and third problems are our next concern.

Clausewitz's conception of nonprescriptive strategic theory consists of two major elements. The first is the nature of the capability of an effective commander-in-chief, which Clausewitz calls "genius." The second is the theoretical apparatus by which the ineffable character of genius can become known to those without experience of high command in war. Aron recognizes the importance of the first element and describes it with remarkable accuracy and perceptiveness. In his early work, Aron observes, the basis of Clausewitz's objective of being "radically antidogmatic" is the formulation of a "theory which indicates that the responsibility for decision is thrown back on the man of action" (ibid., 55). This focus on the commander-in-chief, he understands, is sustained in *On War.* Of the three moral elements of the famous trinity that conclude the first chapter of Book I—emotion, courage and talent, and reason, corresponding to the people, the commander, and the king—it is "that of the commander," Aron says, "which is analyzed at the greatest length." And it is the genius of the commander, he maintains, that "embodies the duality intrinsic to military action." "It alone unites apparently opposed qualities," he writes; "it alone resolves the problems whose complexity baffles the greatest minds and thus, in spite of everything, it presents one of the great and noble expressions of humanity" (ibid., 132–133).

Aron describes Clausewitz's representation of the nature of the difficulties that confront the leader of an army as follows:

The commander must grasp at a stroke, by a kind of intuition, the
truth in the midst of an extraordinary confusion of possibly
contradictory facts. If a Newton himself would have flinched before
such a calculation, it is because the indeterminateness of the quantities
and the singularity of each situation forbid calculation, in the strict
meaning of the word. . . . Genius (or good sense) is exercised less in
the calculation of that which does not lend itself to calculation than in
the discernment of the essential. (Ibid., 185)

In addition, Aron observes, Clausewitz knew that

in war intelligence must triumph over danger, physical effort,
uncertainty and chance. One could just as well say that emotionality
must triumph over its four enemies. In fact, understanding only
triumphs if enlightened by emotion. (Ibid., 135)

Thus Clausewitz, Aron maintains, argues that "the genius of war requires other
qualities than those of the scholar and the scientist, qualities of will and of character which, moreover, only blossom when coaxed by the light of understanding" (ibid., 113).

Aron is unable, however, to provide a satisfactory explanation of Clausewitz's
method of using understanding to coax the development of will and character. He
does ask the right question: "Since rational thought rejects the simplifications of
dogmatic instruction yet seeks to be of use in action, which instruction does it
suggest should be followed?" (ibid., 195).[9] And Aron makes general remarks that
could have served as prologue to sound and specific conclusions. A theory of practice, he observes, "cannot provide a recipe, it must educate the mind" (ibid., 182).
Furthermore, he suggests "that theory does not teach commanders the rules of
war as an art, but only attempts to analyse war as an object" (ibid., 205). Analysis
for Aron means the evaluation of historical examples

with the help of major concepts: centers of gravity, destruction of
enemy forces, strength of the defensive, attrition of the offensive, the
total character of the battle or of the campaign, etc. Compared with
the maxims of the military theoreticians of his time or of ours, these
propositions are of such abstract character that they seem to belong to
another world, and they earned Clausewitz his reputation as a
philosopher or metaphysician. Now, he considers himself above all an
analyst or observer: with good reason. (Ibid., 195)

Aron recognizes that Book II differs from all other sections of *On War* "because it contains a kind of methodological or epistemological commentary on the whole work" (ibid., 177).[10] What he fails to understand is that Clausewitz's discussion of the limitations of conventional military history is accompanied by his solution, the essence of which is that principles not only informed the analysis of outcomes, but shaped the construction of narrative. Conventional operational history, Clausewitz argues in Book II, is incapable of providing an adequate basis for the study of action by the commander-in-chief largely because the historical record does not contain enough information to evaluate the motives that underlay high-level decision-making during crises. For this reason, Clausewitz formulated a theory—based on his own direct observation and personal knowledge of the exercise of high command in war—that identifies and considers the multiple vectors that influence decision-making under difficult conditions, in effect delineating the psychological as well as the material conditions of directing an army on campaign. Productive case study of supreme operational command means the examination of an event described by a combination of verifiable information, which is the property of conventional military history, and intelligent surmise, which can be generated by informed theory. Clausewitz's approach, which in effect reenacts the historical decision-making process, thus eliminates the gap between theory and reality; theory is an integral part of the reconstructed reality being observed, not just an external instrument of analysis.

The objective of reenacting process is not to learn correct procedures, but to engage and come to terms with the rational and emotional elements of command dilemma. Because such conflicts are inherent to any war that poses difficult choices, the applicability of Clausewitzian theory is largely independent of changes in historical circumstances, which, however great, have yet to eliminate the necessity of making hard decisions. Aron, therefore, is incorrect to fault Clausewitzian theory as lacking "any explicit discrimination between historical and transhistorical criteria" (ibid., 143). Moreover, by failing to identify the essential nature of Clausewitzian theory, Aron is reduced to characterizing it in overly general and even banal terms. "Thus the theoretical revolution Clausewitz had in mind," he argues, "was meant not only to discriminate between analysis and doctrine, or permanent data and historical diversity, but also to demonstrate how to use principles based on the majority of cases: neither dogmatism nor heedless improvisation" (ibid., 223). Aron's conclusions elsewhere are no more illuminating. He declares that

> Clausewitz was theorist of an art to be cultivated by study and
> reflection, one that cannot be learned. He inspired, rather than
> instructed. He sought to include everything that pertained to the
> business of fighting in a coherent whole, and to combine abstractions,
> the application of which would vary with circumstances, with a highly
> detailed analysis of the particular circumstances, denying the existence
> of hard-and-fast rules, but suggesting guidelines and maxims,
> qualified by exceptions. (Ibid., 237–238)

In the first chapter of Book I, Clausewitz introduces the argument that defense is a stronger form of war than attack, and he develops this concept in the chapters and books that follow, culminating in a detailed exposition in Book VI, the longest of the eight books of *On War*. Here, Clausewitz presents arguments in support of the superiority of the defense over the attack at both the tactical and strategic level, but his main concern is the latter. He is convinced that, at the strategic level, a willingness to trade space for time on the part of the defender will weaken the attacker—through operational wear and tear, internal political dissension promoted by the lack of immediate success, and logistical overextension—and enable the defender to mobilize popular support, other latent national assets, and the assistance of allies fearful of the dangerous international implications of the attacker's success. These factors in combination will, in Clausewitz's view, be sufficient either to discourage the attacker, and thus limit or even prevent territorial loss, or to garner the defender an opportunity for devastating counterattack, which can provide the platform for positive gains.

In 1976, Clausewitz's views on defense were not generally understood. Liddell Hart's characterization of Clausewitz as the great proponent of the attack was widely known if not universally accepted. The representation of Clausewitz's views on defense in the commentary by Roger Ashley Leonard, on short excerpts from *On War*, published in 1967, was accurate, but too brief to demonstrate their full significance.[11] The essay by Anatol Rapoport introducing the standard abridged English-language edition of *On War*, which appeared in 1968, did not discuss Clausewitz's views on defense, and the abridgement itself omitted all of Book VI.[12] Aron's long and detailed assessment of Clausewitz's views on defense in general, and the exposition of his ideas in Book VI in particular, thus broke new and important ground.

Aron recognized that the relationship between attack and defense was a theme "central to Clausewitzian thought." For this reason, he considered Book VI to be one of the most important in *On War*. Aron was convinced, however, that its meaning was unclear because of its supposedly incomplete state. This belief was based upon Clausewitz's warning, in the note some have dated as having been written in 1830, the year before his death, that Book VI was "only a sketch" and that he "intended to rewrite it entirely and to try to find a solution along other lines." But Aron was troubled by the fact that Book VI seemed to be much more than a preliminary draft (*Clausewitz*, 145, 164).

"Can one call thirty chapters," he observes, "which altogether take up more than 200 pages and which number amongst the most developed of the Treatise, a sketch?" No one, Aron declares, had yet "devoted a rigorous and detailed study" of Book VI. He thus applies an entire chapter to three tasks: that of stating "the guiding ideas of Book VI as they appear to me"; that of finding "faults which become apparent from the text itself, in relation to Clausewitz's requirements of analysis and exposition"; and that of examining "whether the revision, as far as we can predict, would have involved the order, the putting into shape or the modification of some of the principal arguments" (ibid., 145–146).

Aron's rendition of the main arguments of Book VI conform in the main to the description given at the beginning of this section, although he does not address the important issue of the propensity of the attack to be weakened by internal dissension in the event of a protracted war. He notes the presence of the proposition about defense being the stronger form of war in the first chapter of Book I, the one section of *On War* that he regards as having embodied Clausewitz's latest thoughts. This prompts Aron to conclude that the development of the military dimensions of this argument in Book VI would not have changed even if Clausewitz had had more time to work on it. But Aron has three main criticisms of Clausewitz's thinking on this matter. First, he says, Clausewitz fails to connect his examination of the greater strength of the defense either to the question of limited versus unlimited war or to the influence of war on policy. Second, Clausewitz's conceptual scheme is unsatisfactory because it does not provide an adequate explanation of the conditions governing recourse to two different forms of defense, parrying enemy attack as the prelude to counterattack, or parrying enemy attack simply to preserve territory or the army. And third, Clausewitz does not adequately connect his concepts of defense and attack to the trinity of passion, courage and talent, and rationality. Clausewitz, he surmises, would have rectified these flaws had he lived to complete the work (ibid., 163–171).

In general, Aron's complaints are based on his belief that strategic theory is supposed to provide comprehensive explanation of all-important aspects of a phenomenon in order to support productive, if not positive, instruction. Such an approach requires clear statements of terms, propositions, and their relationships. As stated previously, however, Clausewitz rejects this concept of theory in favor of one of his own invention, namely, a theory of observing historical events more accurately than is usually the case, whereby instruction is provided by the act of observing and subsequent analysis, not the theory itself. This concept of theory does not require the apparatus demanded by Aron to achieve its objectives (*On War*, VI/27: 486; VI/30: 516–517). The putative faults enumerated by Aron therefore are not artifacts of incomplete or confused thought, but the intended products of Clausewitz's methods, the nature of which Aron did not recognize.

This fundamental defect of Aron's analysis is abetted by his failure to address critically important statements in the eighth chapter of Book VI that do the very thing he claims Clausewitz did not do: connect the proposition that defense is the stronger form of war to the relationship between war and politics. Here, Clausewitz declares not only that war and politics are connected, but that the attacker is more affected by politics than the defender. For Clausewitz, the attenuation of attacker will, through the debilitating influence of political considerations, is a key issue with respect to the greater strength of the defense (ibid., VI/8: 387). A government corrupted by ignorance and internal division is likely to be even more susceptible to discouragement generated by unfavorable military events. Clausewitz also points out, in the sixth chapter of Book VIII, that while governments are ideally well-informed and united, this is not necessarily or even normally the case (ibid., VIII/6: 606–607).

In his second volume—and thus not in his main analysis of the meaning of *On War*—Aron concedes that Clausewitz defines policy in a way that shows he recognizes its susceptibility to the play of irrational forces. But he suggests that this did not inform Clausewitz's major arguments, and he does not discuss its implications (*Clausewitz*, 266–267). Moreover, in the second volume Aron retails a no less faulty assessment of Clausewitz's views on defense with respect to guerrilla war. Clausewitz, he correctly notes, "preferred defensive, patriotic war to wars of conquest," and "strategically, he liked the defensive, provided it was a just one" (ibid., 237). But his contention that Clausewitz hoped "to avoid the pitiless cruelty that occurs when all are armed and guerrilla warfare is waged by peoples themselves" is not supported by the relevant text in Book VI (*On War*,

VI/26: 479). This analytical error was most likely prompted by Aron's desire to portray his subject as "a preacher of moderation, not excess" (*Clausewitz*, 237). Clausewitz, however, is not predisposed to favor either course, or even to recognize them as relevant attitudes.

It is simply a matter of fact that Clausewitz considers the question of defense in terms of the relationship between war and politics as supplied in the first chapter of Book I, and that he accepts the potential extremes of violence inherent to People's War as in the nature of things. The former observation indicates, as Azar Gat has argued, that Clausewitz revised Book VI *after* composing the note that was supposedly written in 1830. In the note, Clausewitz confessed that his manuscript required a complete revision along fundamentally different lines (*On War*, 70–71). If Gat's contention that this note was misdated, and that it was probably written in 1827 or even earlier, is correct, it means that Clausewitz would have had the time—that is, several years rather than several months—to make major revisions in the manuscript rectifying the shortcomings described in the dated note of July 1827.[13] Aron's approach to Clausewitz was almost certainly based on a misapprehension—namely, the misdating of a note and the conjecture that the bulk of the text of *On War* did not represent its author's latest views. This misapprehension provides an explanation for why Aron would overlook critical sections of Book VI. It is undoubtedly the reason why so much of his analytical scheme is misdirected to surmise what Clausewitz might have written had he lived longer.

Aron's failure to identify the main argument of Book II requires some further explanation. As we will see later in this chapter, W. B. Gallie recognizes that Clausewitz's method of combining history and theory anticipated the work of R. G. Collingwood. Collingwood's concept of history as reenactment, appearing in work he published in the 1920s, is explained clearly in his autobiography of 1939. It is an important argument in his seminal *The Idea of History*, which appeared in 1946.[14] Although Aron's survey of work in the philosophy of history was published in 1938, and in a revised edition published in 1948, he appears not to have been familiar with Collingwood's thinking.[15] And although Collingwood's work received considerable critical attention in the 1950s and 1960s,[16] Aron does not seem to have been aware of it, or if he was, to have recognized its significance with respect to Clausewitz. He thus approached *On War* with a conception of the relationship between history and theory that did not reflect command of scholarship in an area of knowledge that he could claim substantial expertise.

* * *

Aron's direct critique of B. H. Liddell Hart, which opens the second volume of his book on Clausewitz, in effect summarizes his general conclusions about why Clausewitz had been so badly misunderstood. He begins by stating that Liddell Hart is "too intelligent to fail to see that Clausewitz was greater than his disciples, and too English to devote months to unraveling his skein of logical and empirical propositions, of theory and doctrine." Aron then makes four observations. First, he says that an "attack on Clausewitz requires more attentive reading of the treatise, and less summary historical analysis." Second, he states that "you can find what you want to find in the Treatise: all that you need is a selection of quotations, supported by personal prejudice." Third, "If the Treatise is read carefully, the conclusions that it reveals are contrary to those normally drawn from it." And fourth, Clausewitz "does not approve or disapprove, he merely takes note." To these four admonitions he adds remarks that amount to an assertion that Clausewitz is a difficult read because of the contradictions between the first chapter of *On War* and what followed.[17]

On one hand, Raymond Aron was in certain respects well equipped to engage the Clausewitzian mind with success. He read German well, had substantial experience analyzing the philosophical literature of Clausewitz's time, had life experiences that mirrored those of Clausewitz in a way that fostered genuine empathy with his subject, and, like Clausewitz, was a serious student of war and politics. These assets magnified the effects of conscientious reading and vigorous analysis. For these reasons, his examination of *On War* contains much of enduring value.[18] On the other hand, Aron had begun his study with certain debilitating preconceived notions, and he evaluated Clausewitz's writing according to criteria for a satisfactory theory of war that were ill-suited to the object being studied. Although he achieved a great deal, he failed to produce a satisfactory assessment of Clausewitz's work. "Whether one will or no," Aron himself laments, "the teaching of Clausewitz remains and will always remain ambiguous."[19]

Context: Peter Paret

Peter Paret (1924–) was born in Berlin. His family left Germany and, after temporary residence in France and Great Britain, settled in the United States in 1942. Paret served in the United States Army during World War II in the Pacific.

He received his undergraduate degree from the University of California at Berkeley in 1949 and his doctorate from the University of London in 1960. From 1960 to 1962, Paret was a research associate at the Center of International Studies at Princeton University. From 1962 to 1968, he taught at the University of California, Davis, and from 1969 to 1986, at Stanford University. From 1986 to 1997, Paret was the Andrew W. Mellon Professor in the Humanities at the Institute for Advanced Study, Princeton University. While at Princeton in the early 1960s, he helped initiate what became known as the "Clausewitz Project," an ambitious scheme to make Clausewitz's writing more accessible to an English-speaking audience through the translation of many works that were still available only in German. Formally inaugurated by Princeton University in the spring of 1964, the Clausewitz Project ultimately foundered for lack of funding. Paret nonetheless forged ahead with his own scholarship.[20]

Paret wrote about Clausewitz's views on guerrilla war in a brief coauthored study published in 1962. His bibliographical survey of Clausewitz's writing and his first article on Clausewitz both appeared in 1965. Paret published his first major book, a study of military reform in Prussia after its defeat by Napoleon, in 1966. An encyclopedia entry and second article on Clausewitz followed in 1968. Paret's books and articles served as points of departure for his second major book, entitled *Clausewitz and the State,* and for his critical introduction to the new translation of *On War,* both published in 1976.[21]

Paret's work on Clausewitz during the 1960s and 1970s spanned a period of great political turmoil in the United States. An unpopular war, long-standing racial injustice, and the conflicting worldviews of young adults and their parents fomented varying degrees of political controversy and civil disorder in most major American universities. A central issue of this time was the conflict between the moral beliefs that constituted much of what can be called personal identity, on the one hand, and the sometimes acutely immoral behavior of the state, to which individuals nominally owed allegiance, on the other. This may have influenced Paret's formulation of his approach to Clausewitz, whose intellectual development was shaped by similar concerns generated by events that resembled those of the 1960s insofar as they were about fundamental changes in politics and social organization, and the perception that these would shape national destiny in critical ways.

In *Clausewitz and the State,* Paret provides an intellectual biography of Clausewitz based upon comprehensive and rigorous research about the military theorist and command of the intellectual and political history of Germany in the late eighteenth and early nineteenth centuries. Paret contends that for Clausewitz, "the state and its basis, the political vigor of society, occupied places very near the center of his thought." In particular, Clausewitz spent much of his life grappling with the conflict that existed between his ideal of the state as the expression of a national society and the reality of a Prussian state controlled by parochial interests. For this reason, Paret informs his readers, "the interaction between realism and idealism in his life and work constitutes a major theme of this book." Paret observes that any attempt to understand Clausewitz's personal development and political and military actions would be "pointless unless Clausewitz's writings are brought to bear on his life." That said, he states that it is "not the interpretation of Clausewitz's theories but their psychological and historical genesis" that make up his "central theme." Paret also makes it clear that he has "written this book neither to evaluate the adequacy of Clausewitz's theories nor to trace their impact on the conduct of war in the 19th and 20th centuries." Moreover, he admits, "in many respects what I have written is little more than a preliminary outline, which may help others go farther."[22]

Paret's claims and disclaimers suggest that his fundamental difficulty was that a satisfying and therefore generally accepted explanation of the meaning of *On War* did not exist. It was therefore dangerous to attempt an examination of the context, personal and historical, of the many ideas in Clausewitz's greatest work, especially given the high degree of uncertainty about what he meant precisely and the degree of coherence—or noncoherence—of the separate parts. Paret finesses this obstacle by making the object of contextual study not *On War*, but what he regards as Clausewitz's central political concern, namely, his attitude toward the state. By so doing, he could consider Clausewitz's development as an educator, a historian, and a theorist despite the lack of consensus about the views expressed in *On War*. Although one must evaluate Clausewitz's writing to explain his attitudes toward the state, the basic integrity of such an inquiry does not require complete or even wholly valid interpretation of the texts. And a thorough and rigorous examination of Clausewitz's intellectual development in terms of his core political concerns could not help but advance knowledge in ways that offered the prospect of improving understanding of *On War*.

"The state, and the national community which it should represent," Paret observes of Clausewitz, stood for him in a relationship of mutual obligation, a connection that was so close in his mind that he frequently used the term "nation" in referring not only to a people but also to its government and administrative structure. That his concept of the state not only affected his interpretation of political intercourse, but was also linked to very personal feelings further increased its complexity. (*Clausewitz and the State*, 95)

Clausewitz, Paret argues, invests the Prussian state with a moral mission: the preservation of the nation through military efficiency in the face of the dire threat posed by French military power that had been magnified by revolution. The shattering Prussian defeat in 1806 at the hands of the French, however, demonstrated that the existing traditional state and its military apparatus were not capable of accomplishing this task. According to Paret,

When the fighting ended in disaster for Prussia, Clausewitz was at a loss. His recovery, after a period of extreme stress, took two forms: a further intensification of his concept of the state as a heroic autonomous being in the political world, paralleled by an energetic, often reckless effort to bring Prussia closer to that ideal. (Ibid., 119)

The remedy, for Clausewitz, was military reform, which meant not only changes in operational practice by professional soldiers, but national conscription and even resort to national insurrection based upon guerrilla warfare waged by a mobilized citizenry. The objective of improving military effectiveness included the recovery of sufficient military power to free Prussia from French dominance. In early 1812, however, it became clear that the Prussian government would join France in an invasion of Russia. The ratification of the Franco-Prussian alliance prompted Clausewitz to leave the Prussian army and enter Russian military service. He served at first on the Russian general staff, and later with one of the Russian armies that pursued the French during their winter retreat. Clausewitz's actions were a reflection of his belief that his loyalty to the Prussian nation required resistance to France even if it meant disobedience to his king. "In Clausewitz's life," Paret notes, "1812 marks the high point of his open allegiance to a political ideal, rather than to the reality of the politically inadequate Prussian State" (ibid., 232).

Clausewitz fought as a Prussian in the service of Russia from 1812 through 1814. In 1812, he played an instrumental role in negotiations that set the stage for

Prussia's abandonment of the French alliance and reentry into the war as part of an anti-French coalition (ibid., 229–230). In early 1815, Clausewitz was allowed to rejoin the Prussian army. That he could do so was indicative of the separation of his personal ideals from his conception of the Prussian state. As a young man, Paret observes, "he was willing for himself and for the state to take the greatest risks—in part to rectify the historical and human injustices of Napoleon's domination of Germany, but also because at that time he attributed moral and psychological qualities to the state, which, all evidence to the contrary, he insisted must personify heroism and justice." Experience in war convinced Clausewitz that such an outlook was untenable. He "continued to be a Prussian and German patriot," Paret maintains, but he "learned to accept state and people for what they are—suffused with flaws and shortcomings of all kinds" (ibid., 421). "His criticism of Prussian policies," Paret also notes, "was to be extensive [in postwar years] but its tone gradually became very different, less emotional, increasingly objective, at times almost cold" (ibid., 256).

During the war, military reform had been necessary to defeat the French. Following the definitive end of hostilities in 1815, however, conservative antireform forces gained the upper hand. With little prospect of doing constructive work through progressive changes in army organization, Clausewitz agreed to serve as the administrative head of the War College in Berlin in 1818. In March 1819, three months after assuming his post, Clausewitz submitted a critique of the educational program of the War College to Hermann von Boyen, the minister of war, a reformer and an old friend. Clausewitz believed that the curriculum was haphazard, burdened with unnecessary subjects such as literature, deficient in courses on practical military subjects, too focused on rote learning over methods that would foster the development of independent judgment, and satisfied with a form of historical study that was defective in principle. Von Boyen was sympathetic, but general political circumstances—especially increased animus against reform of any kind—ruled out the possibility of a positive response to Clausewitz's proposals. By the end of the year, von Boyen himself had been forced from office (ibid., 272–281).

On the surface, Clausewitz reacted passively to the failure of his attempt to reform the War College and even to the general decline of the reform cause. He did not press his case, and subsequently confined his public activity to administration while devoting his main attention to the composition of *On War* and various historical monographs. Clausewitz's forfeiture of the educational battleground, and redirection of his energy into sustained and serious writing, was not an admission of defeat, however, but a tactical retreat. However unpopular his opinions were among those of his era, the articulation of sound views on the study of war, he was convinced, might find a receptive Prussian audience in the future (ibid., 282–283). Prussian military capability, thereby improved, might make a difference in the war for national survival that Clausewitz believed was bound to occur (ibid., 297). The writing of *On War,* Paret writes, was thus yet another expression of Clausewitz's desire to improve the state's capacity to serve as custodian of the vital interests of the nation. He concludes that Clausewitz's life "demonstrates a unity of motives and effort, a harmonizing of inner needs and achievements, a mastery of reality through understanding" (ibid., 440).

By demonstrating the connection between Clausewitz's intellectual development and his conception of the Prussian state, Paret identifies the motive that drove his writing. This was the rigorous and relentless pursuit of political and military truths, not for their own sake, but in order to promote the survival of his nation through the enhancement of the power of the state. Paret's main line of inquiry is supported by his examination of the development of Clausewitz's views on education, which constituted the instrument by which political and military truth was to be propagated and thus put into practice, and the intellectual sources of Clausewitz's thinking about the state and about education. In his well-balanced and integrated treatment of these issues, Paret explains why Clausewitz wrote *On War* when he did, why he had such little interest in immediate publication, and what, in a general sense, he hoped the work would accomplish. These were significant advances in Clausewitzian studies. Paret's approach to Clausewitz's politics does not, however, provide the basis for a satisfactory explanation of the particular arguments of *On War,* the great majority of which are devoted to fighting in general, and strategy and the exercise of supreme command in particular.

Paret is well aware of the fact that Clausewitz's writing is about much more than politics and that by focusing on his political development he begs important questions about Clausewitz's thoughts on armed conflict. Political and social concepts, he admits, "only dimly illuminate the actual mechanics of war." He justifies giving relatively short shrift to the analysis of the dynamics of fighting by stating that much of the lasting value of *On War* depends on the connection Clausewitz forged between war and politics. In *On War,* Paret argues, "Clausewitz developed a technique of inquiry that sought to identify and separate the numerous components of military organization, decision-making, and action, and to reduce each to its essential core before fitting them together again into larger and dynamic structures." These larger and dynamic structures are political and social, and it is this integration of the military with the political and social, Paret maintains, that "gave Clausewitz's theories their strength and, despite their many transitory aspects, helped make them stimulating to later generations" (ibid., 8).

Paret makes recommendations on how to read *On War* that also favor Clausewitz's handling of the relationship between war and politics over his treatment of the dynamics of war. On the strength of the notes of 1827 and 1830, Paret believes that the first chapter of Book I is the only chapter to have been completed to Clausewitz's satisfaction, and that as such it indicates the direction that Clausewitz intended to follow in the later sections of *On War*. Paret is convinced, in particular, that its arguments dealing with limited war, and the concept that war is an extension of politics, were to be the basis of a revision of *On War* that was planned but never executed. He thus argues that "the reader of *On War* will find himself in accord with its author if he gives the political motives and character of war more prominence than they receive in much of the text, and, further, if he amends the unrevised sections to the effect that limited wars need not be a modification, but that theoretically as well as in reality two equally valid types of war exist." To facilitate such action, Paret reproduces the entirety of the first chapter of Book I in the form of a complete subsection of the chapter on major historical and theoretical works (ibid., 379, 382–395).

Paret's use of the first chapter of Book I as the basis of interpreting the balance of *On War* is similar to Aron's approach. As I noted previously, in the section on Aron, such a course is problematical; here, a more elaborate discussion of the problem can set the stage for a detailed consideration of the weaknesses of Paret's analysis of *On War*. As stated previously, it is likely that the undated note of 1830 was actually produced in 1827, and therefore, there is a strong possibility that *On War* was revised to address the issues raised in the dated note of July 1827. If this is the case, then the text of *On War* in its entirety should more or less accurately represent Clausewitz's late views on the relationship of war and politics, limited war, and other matters. To give these two issues greater weight than is warranted by the text, therefore, is to distort Clausewitz's exposition. In addition, Clausewitz concludes the dated note of 1827 with the observation that "an unprejudiced reader" of the first six books of *On War,* as it stood, might be able to discern "the basic ideas that might bring about a revolution in the theory of war." These words indicate that Clausewitz believed that the *entirety* of his manuscript, including even the incomplete and unrevised portions, revealed propositions that challenged existing theories in fundamental ways, and that by an open-minded reading the discerning reader could grasp these concepts.

Paret believes that Clausewitz did not intend much by this remark. Its author, he argues, "may have meant no more than to express his confidence that *On War* indicated what questions ought to be asked, and the kind of answers, combining specificity with universality, that should be sought" (ibid., 381). But Clausewitz's claim to have expressed transformative views even in his incomplete and unrevised redaction cannot be so easily dismissed. As mentioned earlier, the text of *On War* presents two lines of argument, distinguishable from those on the relationship between war and politics and on the two forms of war, that could be considered revolutionary. These are, first, Clausewitz's contention that defense is a stronger form of war than attack, and second, his description of a method of combining history and theory in a way that more accurately represents the nature of command dilemma than conventional military history. Paret pays little attention to Clausewitz's views on the defense, and his extensive discussion of Clausewitz's historical and theoretical writing do not do justice to the manner in which the two are integrated in *On War*. Both issues, as well as other aspects of Clausewitz's work that Paret does not address, require careful consideration.

Clausewitz presents his main arguments on defense as the stronger form of war in Book VI of *On War*. This section is more than triple the length of Book I and, as explained previously, contains important material that clearly reflects Clausewitz's late views on the relationship between war and politics. Aron justifiably regards it as one of the most significant portions of *On War,* and furthermore believes that Clausewitz's ideas about defensive war are central to his theoretical concerns. Paret, in contrast, refers to the argument that the defense is the stronger form of war only in passing, in a passage describing the undated note of 1830. He does not engage the argument, and indeed there is not a single reference to Book VI in the portion of Paret's book that contains his formal analysis of Clausewitz's theories. Clausewitz favors the defensive even in his historical accounts of campaigns in which the attacker was extraordinarily successful, and Paret's reactions to this emphasis are revealing. He does not comment on Clausewitz's belief that Prussia could have avoided disaster in 1806 through

a defensive strategy based on a holding action followed by major counterattack, for example, consigning even a mention of this idea to a footnote (ibid.,
345). And Paret finds it "startling" that Clausewitz would attempt to illustrate
the potential strength of the defensive by bringing up the allied attack on
France in 1814, which Paret regards as "one of the most successful offensives in
the history of war" (ibid., 359).

Clausewitz presents his conception of the interrelationship of theory and
history in Book II, calling for the augmentation of verifiable evidence about a
past military event by means of a theoretical model to produce an account that
mixes history and theory-based surmise. Such a hybrid of history and theory,
Clausewitz believed, is a better representation of reality than can be achieved
through conventional historical methods, thus providing a sounder object of
study for the purposes of critical analysis. Paret devotes much less attention to
Book II than Aron does, and like Aron appears to have been unfamiliar with the
work of R. G. Collingwood, whose outlook on how to engage the past is similar
to that of Clausewitz. It is thus not surprising that Paret, like Aron, was unable
to recognize the role of what is in effect historical reenactment in Clausewitzian theory. Instead of identifying Clausewitz's unification of history and theory, Paret depicts the relationship of history and theory as reciprocally interactive, with the former used to modulate the prescriptive tendencies of the latter
by illustrating the degree to which all theory was subject to qualification, and
the latter used to augment the narrative of the former with critical argument
(ibid., 328–329).

On the important question of the nature of policy, Paret misrepresents
Clausewitz's position. In Chapter 6 of Book VIII, Clausewitz declares that policy
is not always the product of rational political considerations. In so doing, he
characterizes the influence of corrupting factors, such as personal ambition or
private interest, as neither good nor bad, but simply in the nature of things (*On
War*, VIII/6: 606–607). Paret, however, provides a partial quotation of the passage in question—taking Clausewitz's words out of context—to support his
contention that Clausewitz believed political purpose was "generally realistic
and responsible," even though he recognizes that this is not the position Clausewitz took in his political essays (*Clausewitz and the State*, 369). Thus Paret, like
Aron, oversimplifies Clausewitz's conception of policy by removing political
friction from the field of observation, thereby rendering Clausewitz's notion of
politics in ideal rather than realistic terms. Such an interpretation obviously
would have been unacceptable to Clausewitz.

Paret's fundamental problem with respect to *On War* is that he misunderstands Clausewitz's theoretical intent, which is to say that he, like Aron, believes that Clausewitzian theory put forward a set of propositions devoted to the explanation of the essential nature of war (ibid., 357–358, 365). On the basis of this supposition, he evaluates Clausewitz in terms of the completeness of his perspective and the soundness of his particular analyses. He is therefore critical of what he perceives as omissions: Clausewitz's failure to address naval warfare, the ethics of violence, the danger posed by irrational political leadership, and the possibility that excessive military mobilization can have counterproductive effects. Moreover, he criticizes the conventional character of Clausewitz's taxonomy of war. "The weakest part of *On War*," Paret says, is Clausewitz's discussion of psychological elements (ibid., 365–377).

Clausewitz's primary theoretical concern, however, is the integrity of the processes of observation and analysis of past military events. Reaching general conclusions about war as a phenomenon is not the main point; rather, his discussion of certain aspects of war has two functions. First, he sets out to explain why conventional history is an inadequate basis for sound critical analysis and thus to justify the creation of a more suitable object for observation through the combination of historical fact and theoretical surmise; and second, he wants to facilitate the proper study of a case history so constituted. To achieve these objectives, it is not necessary to undertake a comprehensive explanation of war. Clausewitz indeed is convinced that a comprehensive explanation of armed conflict is either impossible to produce in the form of pure theory, or realizable only to a degree through the representation of the past by means of historical case studies. Paret's commentary on *On War*, therefore, can be regarded as not so much wrong as in important respects misdirected.

Paret's overgeneralized and incomplete analysis of Clausewitz's primary arguments about the military dynamics of armed conflict reveals little that can be regarded as significant. As a consequence, by virtual default, his main assessment of *On War* as a work of theoretical importance has to focus on the relationship between war and politics and on the categories of limited and unlimited war. Here, Paret provides a reasonable resolution of the antinomy that had tormented Aron. Clausewitz, he argues,

> was convinced that often in the past limited conflicts had occurred,
> not because the protagonists' means precluded greater effort or
> because their leadership or will had faltered but because their

intentions were too restricted to justify anything more. A war fought
for limited goals was not necessarily a modification or corruption of
the theoretical principle of absolute war. . . . Limited wars might be a
modification of the absolute, but need not be, if the purpose for which
they were waged was also limited. Violence continued to be the
essence, the regulative idea, even of limited wars fought for limited
ends; but in such cases the essence did not require its fullest possible
expression. The concept of absolute war had by no means become
invalid and it continued to perform decisive analytic functions; but it
was now joined by the concept of limited war. (Ibid., 377–378)

Paret maintains that Clausewitz's thinking on limited and unlimited war, as
Paret had interpreted it, constitutes the Prussian author's "most impressive intellectual and psychological achievement" (ibid., 381). The ideas regarded as
most significant by Paret, however, are too abstract, if not platitudinous, to offer
much to practical soldiers. For this reason alone, Clausewitz himself would almost certainly have disagreed.

But Paret's main purpose in *Clausewitz and the State,* it will be recalled, is not to
evaluate Clausewitz's writing, but to explore the sources of his manner of thinking. Paret's conclusions in this area are based on careful and well-informed assessments of Clausewitz's intellectual character and that of his contemporaries.
Paret attributes Clausewitz's dialectical reasoning to conformity with the general practice of his time, and as such says it is not an indicator of allegiance to
any particular school of philosophy. "His argumentation," Paret observes, "was
characterized by the dialectical forms that were the common property of his
generation" (ibid., 150).[23] But Paret maintains that the substance of Clausewitz's
approach to intellectual problem-solving is unique in its eclecticism:

In the great movement of Romantic idealism that dominated German
intellectual life at the turn of the century, Clausewitz's position is
easily recognized. He rejected the popular Enlightenment, with its
doctrinaire faith in rationality and progress, and found no difficulty in
acknowledging limitations to human understanding, an acceptance
that made reason all the stronger in those areas left to it. He benefited
enormously from the liberating emphasis that the early Romantics
placed on the psychological qualities of the individual; but he did not
follow such writers as Novalis or the Schlegel brothers in their

surrender to emotion. The religious wave of Romanticism did not
touch him; nor did its mysticism, nostalgia, and its sham-medieval,
patriarchal view of the state. In feeling and manner he was far closer to
the men who had passed through the anti-rationalist revolt of the
Sturm und Drang to seek internal and external harmony, and who
gave expression to their belief in the unity of all phenomena—"the
marvelously structured organic whole of all living nature," as
Clausewitz once wrote—not in rhapsodic musings after the infinite,
but in disciplined mastery of thought and form. The ideal of antiquity
that inspired Goethe and Schiller had, however, no hold on
Clausewitz; its place was taken by a fascination with political and
military reality. (Ibid., 149–150)

We "cannot point to any single ancestor," Paret thus concludes, "of his theoretical method among German philosophers" (ibid., 150). He states the reason
for this most clearly in the introduction to his book. Clausewitz's conceptual approach, Paret maintains—with good cause—"was as much the result of action
and experience as it was of speculative effort" (ibid., 5).

Paret's findings about more specific aspects of Clausewitz's thinking can be
employed to support interpretations of *On War* that differ from his own. In
Clausewitz's early writing on the requirements of theory, Paret notes, Clausewitz had stated that the "study of psychological forces . . . must move beyond the
marginal acknowledgement of their existence to full integration into the theoretical structure" (ibid., 158). He recognizes that in *On War* Clausewitz articulated views that made the "genius" of the commander-in-chief critical, if not
central, to the theoretical effort. According to Paret, Clausewitz believed that
theorists must take care "not to inhibit the action of genius in their doctrines"
and to afford "scope to freedom of action in general" (ibid., 161). Paret also
comes remarkably close to identifying the general principle underlying
Clausewitz's unification of history and theory. The Prussian, he explains, taking
a "view of history different from that held by most military writers of the time,"
"reversed the functions of history and theory." Military history, for Clausewitz,
was "not a pool of material for the theorist"; instead, one purpose of theory was
"to help us understand the past" (ibid., 198).

Paret recognizes that Clausewitz preferred accurate observation of what
he believed to be reality to theoretical abstraction. He also notes that this practical outlook separates his approach from the academic philosophers of his
time. "In one respect," Paret writes, "Clausewitz differed completely from the

transcendental philosophers," in that "he accepted the reality of concrete phenomena," which are described as "those physical, moral, and psychological
phenomena that were contained in the record of history and that everyday experience presented to us." Paret notes that Clausewitz expressed his hostility
to detached theorizing in a note of 1808 entitled "In reference to well-meaning
German philosophers." Here the Prussian officer argues that "contempt and
derision are merited by presumptuous philosophy, which seeks to raise us
high above the activities of the day, so that we can escape their pressures and
cease all inner resistance to them." "That even realistic theory could never
match reality," Paret concludes, "remained a fact of crucial significance to
Clausewitz, and he never ceased to emphasize the scientific and pedagogic
consequences that flowed from the difference" (ibid., 151, 163).

In his earliest historical writing, Paret observes, Clausewitz sharply criti-
cized conventional military historical scholarship for paying inadequate attention to the psychological factors that played a role in the exercise of high command in war (ibid., 85–90). In his later historical writing, Paret argues,
Clausewitz's concern for the factors that generated dilemmas for generals continued. "Clausewitz," he maintains, "found it easier to reach 'genuine conviction'
on strategy and generalship than on the events of a particular engagement, to
which . . . he ascribed less significance" (ibid., 335). Clausewitz, Paret also states,

> tried to put himself in the position of the people he wrote about,
> although he knew that an approximation was the best that could be
> achieved. His constant discussion of alternatives open to government
> and individuals is one expression of this effort. At the same time he
> sought to discover what almost always lay beyond the knowledge of
> the historical actor: the larger connection of events. (Ibid., 332)

Paret notes that Clausewitz was consistently willing "to give the benefit of
the doubt to men acting under great difficulties," and that, "even under the most
favorable circumstances, the historian would find judging men's behavior nearly
as doubtful an enterprise as offering them instruction: both can easily interfere
with true understanding" (ibid., 340).

Paret came close to making an accurate—if still overly general—assessment
of Clausewitz's approach to the military dimension of war in his comparison of
his work to that of Jomini. Clausewitz, he observes, believed that an "emphasis
on history . . . became the force that more effectively than any other restricted the
authority of principles or rules of war." In essence, this approach acknowledges

"the infinite variety of war, as of life, without surrendering to it" (ibid., 203–204). He then evaluates the concluding chapter of the fourth volume of Jomini's *Treatise on Major Military Operations*. In this work of 1811, Jomini proclaims the universal applicability of what he calls the "fundamental principle"—namely, the combination "of the greatest amount of one's forces against the decisive point"—which is to be achieved through obedience to ten maxims of appropriate conduct. Paret's assessment of Clausewitzian and Jominian methodology favors the former in no uncertain terms:

> Jomini pursued an arbitrary number of facts and observations
> through a disorganized argument to a dogmatic end. Clausewitz
> developed his ideas as logically as he knew how[,] to produce not a
> system but insights into the nature and interaction of military
> phenomena, which enabled the student to progress further through
> the infinite variety of dangers and opportunities. (Ibid., 204–205)

Paret reprises much of what he had argued in *Clausewitz and the State* in his introductory essay to the new translation of *On War*. In "The Genesis of *On War*," he insists that the unfinished state of the book means that it did not reflect the author's latest thinking about the influence of politics on war or about limited and unlimited war. He thus repeats his injunction to give greater emphasis to these matters than they received in the text ("The Genesis of *On War*," 22–23).[24] Paret also characterizes *On War* as an attempt to explain the essential nature of war. Clausewitz's aim, he declares, "was to achieve a logical structuring of reality." Paret emphasizes the importance of the "genius" of the commander-in-chief. The "role of genius," he observes of Clausewitz's treatment of the subject, "lies near the source of his entire theoretical effort" (ibid., 3, 16).

Paret does not address Clausewitz's argument that defense is the stronger form of war. Nor does his treatment of the relationship between history and theory identify Clausewitz's method of augmenting the historical record with theoretical constructions to create a representation of the past that is a closer approximation of reality than is achievable through conventional historical narrative alone. And yet the tone of his language suggests an awareness that Clausewitz had been up to something important that still evaded clear delineation:

> In Clausewitz's pedagogic and theoretical work, history had the
> additional function of expanding the student's or reader's experience,
> or substituting for it when experience was lacking. History depicted

reality and stood for reality. The role of theory, on the contrary, Clausewitz once declared, was merely to help us comprehend history—a highly telling reversal of roles that few other theorists would have agreed with or even understood. (Ibid., 23)

Paret's later representations of the relationship between history and theory in *On War* did not develop significantly beyond those expressed in 1976.[25] This may explain why, even sixteen years after the publication of *Clausewitz and the State,* its author remained uneasy. The "relationship between theory and history in Clausewitz's thought," Paret admitted in 1992, "has been barely studied."[26]

Method: W. B. Gallie

Walter Bryce Gallie (1912–1998) was the third son of a structural engineer. He received his bachelor's degree in 1934 from Balliol College at Oxford and began his academic career in 1935 teaching philosophy at the University College of Swansea. He served in the British army during World War II, at its end holding the rank of major and receiving the Croix de Guerre. Gallie earned his master's degree at Oxford in 1947 and resumed his teaching at Swansea. He moved to the University College of North Staffordshire in 1950 and from 1954 to 1967 served as professor of logic and metaphysics at Queen's University, Belfast. From 1967 to 1978 he was professor of political science and a Fellow of Peterhouse, Cambridge University.[27] Prior to writing his essay on Clausewitz, Gallie published two monographs on subjects that would seem, at first glance, to be unrelated to military affairs, but that nevertheless productively informed his examination of the Prussian author.

Gallie's first scholarly monograph was a study of the American philosopher Charles Sanders Peirce (1839–1914). Entitled *Peirce and Pragmatism,* it was published in 1952. Two important characteristics of Peirce's philosophical method, Gallie argues in this work, are his insistence "that philosophy should imitate the sciences by making its premises and methods more explicit," and that it should establish "generally acceptable standards of correct usage and sufficient proof (with the proviso that no given expression of these standards should be taken as perennially adequate)."[28] Peirce's thought, he observes, "so far from being chaotic (as is sometimes alleged), was in fact too closely interconnected over too wide a field to admit easily of effective literary expression."[29] Such characterizations are applicable to Clausewitz. The Prussian author's determination to formulate a scientific method of studying armed conflict caused him in *On War* to describe his assumptions and manner of reasoning at length, and in particular to formulate criteria for sound study that avoided setting hard and fast rules. Clausewitz's analytical vision, moreover, is, like that of Peirce, well-ordered but nonetheless extremely difficult to comprehend because of its complexity and interconnectedness. Studying Peirce, in short, probably primed Gallie to take on the subject of Clausewitz.

In 1964, Gallie published *Philosophy and the Historical Understanding*. This monograph examines, among other things, the writing of R. G. Collingwood (1889–1943) in *The Idea of History,* which had been published posthumously in 1946. Gallie recognizes that an essential feature of Collingwood's concept of history in this work is his insistence that past events are about problem-solving by human beings and that a valid historical reconstruction of an event requires the reenactment of the thought processes of the people who played critical roles in solving the problems in question. Collingwood, Gallie observes,

> maintains that I can solve an historical problem only when I succeed
> in re-thinking some other man's—say Plato's or Napoleon's—
> thought; and, on the other hand, I can only do this or at least can
> only know that I have done this correctly, when the thought which
> (in re-thinking it) I attribute to Plato or Napoleon has the effect of
> solving the problem, in either Platonic or Napoleonic history, from
> which my original puzzlement, and hence my need to think
> historically, arose. In other words, successful (or the appearance or
> the sensation of successful) problem-solving is taken as a sign of the
> reality, the re-lived-ness, of the thoughts which the historian ascribes
> to his subject.[30]

Gallie's description of Collingwood's view of history as reenactment, and especially his use of Napoleon as an illustrative example, invites comparison of the concept to Clausewitz's call for the recreation of the psychological circumstances of decision-making by the commander-in-chief of an army.

When Gallie gave the Wiles Lectures at Queen's University in May 1976, his presentations were concerned with "the roles and causes of war and the possibilities and conditions of peace between the peoples of the world," matters which he regarded as "the most urgent political problems of our age."[31] In these lectures, he chose to examine the relevant works of Immanuel Kant, Carl von Clausewitz, Karl Marx and Friedrich Engels, and Count Leo Tolstoy as sources of intelligent—and therefore useful—commentary on his chosen theme. Gallie revised and "slightly extended" the texts of his talks, which were then published in 1978 under the title *Philosophers of Peace and War*. In this work, Gallie acknowledged the publication of the new translation of *On War* and the monographs of Aron and Paret, whose contributions to the understanding of Clausewitz he addressed in a general way. His detailed criticism of these works, however, was reserved for an article published in the *European Journal of Sociology* in the same year.[32]

In the Preface to *Philosophers of Peace and War*, Gallie notes that the commentator on his lecture on Clausewitz had informed him of the impending publication of the monographs of Raymond Aron and Peter Paret, and the new translation of *On War* (*Philosophers*, ix). In his Introduction, Gallie singles out Aron's book in particular, praising it "as expectably impressive in its range and accuracy of treatment as it is moving in the generous sympathy of its spirit" (ibid., 6). In a "bibliographical note" he observes that Aron, Paret, and Howard had made contributions to Clausewitzian scholarship comparable to the greatest of their predecessors. Aron had provided "a comprehensive, sympathetic and uniformly wise survey of all the theoretical problems that arise from studying *On War*." Paret "provides the indispensable social background for a just appreciation of it." And the new translation of *On War*, Gallie maintains, "reads much more smoothly" than the previous standard text of Colonel J. J. Graham. Although Gallie was not always in agreement with the new translation, he regarded it generally as "much more reliable." In his footnotes to the chapter on Clausewitz, Gallie cites all three works (ibid., 142–146).

In his Introduction, however, Gallie makes it clear that he believes that in spite of the efforts of Aron, Paret, and Howard, the main arguments of *On War* had yet to be understood. Clausewitz's great work, he argues,

> was left unfinished, and contains some fundamental inconsistencies, and many of its most important ideas are introduced in the most unexpected places, almost as marginalia or asides. Moreover, despite years of effort, Clausewitz never found a satisfactory way of expressing the central insight upon which most of his arguments hinge. His idea of Absolute War has never been adequately expounded, because it has never been adequately analysed and criticized with respect to its origins, to its place in Clausewitz's conceptual system, and to the confusions with which it is enabled in the opening chapter of his book. So Clausewitz, although his style has been compared for elegance with Goethe's and his best sayings belong to world-literature, stands in need of much sympathetic and critical reinterpretation. . . .

Where Clausewitzian scholarship still falls short [Gallie
concludes] is in philosophical appreciation and criticism: neither the
originality of Clausewitz's general philosophy of action nor the logical
confusions involved in the doctrine of Absolute War have as yet
received adequate attention from philosophers." (Ibid., 4, 6)

Clausewitz, Gallie argues, deserves serious philosophical criticism for the sim-
ple reason that the reasoning in *On War* is philosophically sophisticated.
Clausewitz's undated note ("On the Genesis of His Early Manuscript on the
Theory of War," c. 1818), which appeared in a preliminary version of *On War*,
Gallie notes, reveals a philosopher's temperament. Writing of Clausewitz's re-
marks, which are quoted almost in full, he says:

> It seems to me that whether these words were prefixed to a treatise
> on war or a treatise on peace—or on law or on rhetoric or on logic
> or mathematics or economics or navigation—no one with the
> slightest acquaintance with philosophy could fail to suspect that
> their author was a man of marked philosophical ability. They
> display the comprehensive view, the poise, the slightly ironic self-
> awareness, the modesty and the assurance that are necessary to any
> work of value in that field. They are the words of a man who knows
> very well what he is about, and yet also how little—and at best how
> one-sidedly—he can convey that knowledge. Above all they suggest
> a man who realizes to what extent any thinker is in the hands of his
> work, that *his* best work is when *it* takes over, and that his main
> duty is to ensure that this happens as often as possible. Which
> suggests a genuine, if not a great, philosopher. (Ibid., 42; italics in
> original)

Clausewitz, Gallie argues, made three main contributions to the understanding of war in philosophical terms. First, Clausewitz maintains that "the
ideal of a logically complete or sufficient 'answer' to any problem in warfare is
a sheer delusion," because war is a social rather than either a technical or an artistic phenomenon. Second, although principles of war are necessary in a subjective or educational sense, from the standpoint of actual practice they are
never "*sufficient* to decide what must be done in any war-situation," nor are
they "ever *necessary* to a right military decision." And third, Clausewitz focuses
on the action of the commander-in-chief, whose decision-making skill is about

the ability to make correct judgments rather than the possession of objective knowledge (ibid., 43–46; italics in original).

Collectively, these propositions amount to the delineation of a theory of practice that insofar as the early nineteenth century is concerned reflect a novel philosophical outlook. The academic philosophers of Clausewitz's time, Gallie argues, were largely concerned with "the understanding of a mathematical principle or of a fairly complex piece of machinery, or even the mastery of a complex technique." That is to say, the standard approach of philosophers in Clausewitz's day was to formulate objective and therefore impersonal propositions about phenomena. Clausewitz, in contrast, called for the imagination of the highly variable, contingent, fraught, and therefore personal nature of the capacity to respond effectively to difficult situations. Such an approach, Gallie argues, "would have been much appreciated by Aristotle, and, oddly enough, by some of the ablest philosophers of our century," who undoubtedly included Peirce, although he is not referred to by name, and certainly Collingwood. The approach also has, Gallie says, "important implications for human life far beyond the military field" (ibid., 42–43, 46).

Gallie notes that the investigation of congruences of methodological outlook between Clausewitz and advanced philosophical work of the twentieth century would be "interesting." He even goes so far as to state that Clausewitz's concept of historical thinking, as given in Chapter 5, Book II, of *On War*, resembles that of Collingwood. "Clausewitz's discussion of the use of military history, and the critical study of the best documented battles and campaigns," Gallie argues, "anticipates R. G. Collingwood's idea of history as the re-enactment of past deeds, while yet taking account of the 'hard' facts of history, which Collingwood, with his idealist predilections, either minimizes or ignores." Gallie declines, however, to take his explorations further, being satisfied with doing no more than "indicating how original and how strong Clausewitz's philosophical capacities were" (ibid., 43, 46, 145n12). This exercise, he explains, is necessary for two reasons. First, it answers the charge of Basil Liddell Hart and J. F. C. Fuller (1878–1966) that Clausewitz's work is philosophically impoverished.[33] Second, and more importantly, it sets up his main analysis of *On War*.

Despite his admiration for Clausewitz's philosophical talent, Gallie is convinced that the exposition in *On War* is compromised by faulty philosophical methodology. "Clausewitz does very well with his own unprofessional but carefully considered arguments and analyses," he argues, but when he relies "upon what he takes to be accepted philosophical terms and methods . . . his touch fails and . . . he lapses into errors and equivocations." The main location of defective exposition, Gallie maintains, is the first chapter of Book I. Unlike both Aron and Paret, Gallie is highly critical of this section of *On War*. The "conclusions of the revised opening chapter," he writes, "lack the penetrative power—or the promise of penetration and illumination—which is so strongly conveyed in many (relatively) unrevised passages of *On War*" (*Philosophers*, 47–48).

Two other factors also militate against the usefulness of Book I, Chapter 1, as a guide to Clausewitz's main thinking on armed conflict, according to Gallie. In the first place, Gallie believes that if Clausewitz had lived to revise the balance of *On War,* he might well have recognized the weaknesses of this chapter and made major changes to correct them. In the second place, Clausewitz's presentation of arguments about absolute and real war are neither clear nor coherent (ibid., 48). The more the analytical scheme of the opening chapter of *On War* is examined, he concludes,

> the less likely we are to regard it as sacrosanct, or as indispensable for the appreciation of the different and sometimes conflicting strands in his thought. The conceptual framework in question has, to my knowledge, never been adequately unpacked, scrutinized, and reconstructed for comparative purposes; it has never been submitted to systematic logical criticism; and the question of its main philosophical source or inspiration has never even been raised. (Ibid.)

Gallie thus declares that the primary purpose of his own chapter is to rectify the situation just described and provide the foundation for clear comprehension of *On War*:

> I want to show . . . that Clausewitz's conceptual framework, as articulated in Book I chapter I, is fatally flawed; and that until its flaws are understood, and until it is replaced by something much simpler and logically sounder, the unity of his thought must remain blurred— even its most brilliant insights have to be seized as if through a fog of mystification and distortion. (Ibid.)

Gallie's instruments of clarification are philosophically informed investigation and reinterpretation. It "seems to me intolerable," he concludes, that so much of Clausewitz should have remained, for most of his readers, veiled in a semi-intelligibility which a little criticism and reconstruction could easily have dispersed; the more so since his weaknesses and failures are mainly due to his unhappy philosophical, and more specifically his logical, inheritance. (Ibid., 49)

Gallie believes that the analytical scheme put forward in the first chapter of Book I is based upon Clausewitz's descriptions of absolute and real war. Clausewitz, he argues, presents the former as governed by the principle of maximization of violence, and the latter as a phenomenon in which the use of violence is restrained to varying degrees by political considerations. According to Gallie, in Clausewitz's analytical scheme absolute war constituted "the inner logic of the operation of war," whereas real war "revealed its essential social function." Gallie accepts Clausewitz's characterization of real war on the grounds that "there is no problem about the meaning of the terms involved." This is not true, however, of absolute war, which is "a term of art, whose place in his thought Clausewitz nowhere explains as fully as its importance requires" (ibid., 52). Gallie, in other words, holds that while Clausewitz's definition of real war is evidently valid on its face, his assertion that absolute war is defined by a particular principle requires proof.

Before addressing the issue of evidence, Gallie investigates the matter of motive—that is, what prompted Clausewitz to introduce the concept of absolute war in the first place. He believes, unlike Aron and Paret, that Clausewitz was much influenced by Kantian methodology. Clausewitz's dual characterization of war into the categories of absolute and real, he argues, is a manifestation of Kant's principle of division. As Gallie describes it, this means that "any major division within human interests, for instance between formal and empirical knowledge or between the rational and the animal (or mechanical) elements in human conduct. . . . should be stated in the sharpest, most extreme possible form" (ibid.).

In addition, Clausewitz's conviction that the dynamics of absolute war represent what Gallie calls "the most important fact about war" is strengthened by two other factors. In the first place, Gallie observes, it seemed to Clausewitz "that Napoleon owed his success to the fact that he planned his campaigns and battles in ways that approximated closely to the idea of Absolute War"—that is, that "war is a serious business, in which the issues can well be the survival or extinction of nations." And second, "when Clausewitz came to write papers and chapters on military training, he noticed that traditionally acknowledged principles of tactics and strategy are much more clearly and effectively exemplified in wars that approximate to the Absolute form" (ibid., 53).

Having described Clausewitz's motives for advancing absolute war as a major analytical category, Gallie turns to the matter of the logic of its presentation. Clausewitz, according to Gallie, makes two contradictory claims that compromise the theoretical utility of the concept. The first is that absolute war is a necessary truth, and the second is that abstract terms or ideas in general are falsifications of reality because they unavoidably select and simplify, and therefore mislead. Gallie is unwilling to accept the first proposition without proof, which is not provided in the opening chapter of *On War*. As for the second, he responds that all theory involves selection and simplification, and it is thus "pointless to impugn any particular abstraction simply for being abstract; the pertinent question, in all cases, is whether some particular abstraction is or is not helpful, illuminating, testable, adequate, within the limits of a given problematic area." Both of Clausewitz's contentions, he declares, are "very bad arguments" (ibid., 54–55).

The foregoing does not, however, prompt Gallie to reject absolute war as a concept of the first importance. Instead, he finds an alternative way of looking at absolute war that justifies the great significance assigned to it by Clausewitz. Gallie interprets the second chapter of Book I[34] as follows:

> The most finely calculated strategies based on balancing one's own and one's opponent's risks and advantages, may leave a too clever commander exposed to an unexpected knockout blow from an opponent whose resources or daring he has underestimated—the most terrible calamity that can ever happen to a commander in war. It is this danger—the permanent possibility rather than the calculated probability of a knockout—that, more than anything else, explains Clausewitz's obsession with Absolute War. It must always be at the very forefront of a commander's mind because it represents both the very worst and the very best results with which he can be accredited. (*Philosophers*, 58)

Before applying this concept to a reinterpretation of what he regards as the primary arguments of the opening chapter of *On War*, Gallie puts forward a proposition that is not only confusing, because it undercuts his main line of reasoning, but also demonstrably faulty. Clausewitz, he maintains, tended to write "from the point of view of a commander accepting an assignment and deciding, at the outset, how he can best carry it through—in effect deciding on the lines and character of the coming battle or campaign." Gallie concedes that Clausewitz showed "some appreciation" of the degree to which an opponent's action might force modification of the initial strategic plan, but he maintains that Clausewitz only revealed this "indirectly" through his insistence "on the importance, for every commander, of persisting in his assigned objective against the temptation to veer from it as circumstances, danger or opportunity may suggest." Clausewitz, he concludes, did "not appreciate . . . that in the course of a campaign, even the most resolute commander may be forced, on pain of the total destruction of his army, to move from one form (aim and strategy) of war to another" (ibid., 59).

This characterization of Clausewitz's thinking is mistaken. The very opposite position is presented in no uncertain terms in the second chapter of Book I (*On War*, I/2: 92) and developed with even greater force in the eighth chapter of Book VI. Gallie's extended discussion of Clausewitzian error is thus unnecessary; moreover, his sports analogy, offered as a corrective, in fact represents Clausewitz's actual views. The analogy compares war to boxing:

> Consider . . . an ordinary boxing bout. It is quite possible that each boxer will start with a clear and definite plan in mind: he will aim at victory on points or at victory by knockout, and, after a little preliminary manoeuvring, will proceed accordingly. But notoriously, a contestant who begins with one plan in mind may be led by opportunity, or by desperation in the face of unexpected resistance, or by the sheer crescendo of the struggle, to change his method of fighting in the course of the bout. The threat of a certain points-defeat may drive him into frantic efforts to score a knockout, or the impossibility of securing a knockout may induce him to fight for a bare victory, or a draw, or even an honourable defeat on points. Thus every act of fighting, and not only war, contains in itself the seeds or the possibility of either type of result; and the nature of the result, quite as much as the question of who will be victor, may remain in doubt until the very end of the contest. (*Philosophers*, 59–60)

Gallie concludes with a summation of his "proposed reconstruction of Clausewitz's conceptual system." This consists of two strands. The first is a representation of war as a phenomenon in which the "logic of fighting" is subordinated "entirely to the idea of War as a Political Instrument." The second is that there are "two fundamentally different ways of fighting—one in which every exertion is directed to the knockout blow, the other in which position, strength and resources are exploited to obtain any available advantage." "Emphasizing the frequent transitions between the two kinds of war," Gallie maintains,

> . . . enables us to do justice to both the following propositions: first that, in historical truth, very few wars come near to the Absolute form, and yet, secondly, that the mere possibility of a war approximating to the Absolute form is and ought to be the predominant thought in every commander's mind, since it represents either supreme success or the worst disaster that can possibly befall him.

Gallie declares that such a characterization of Clausewitzian theory has two major advantages: "that it does reconcile, in the sense of doing equal justice to, the two main pulls (and poles) of his military thought; and that it is a position which accords well with his historical vision of war—of the development of war that he had witnessed and of the wholly incalculable future which he ascribed to it" (ibid., 60).

Gallie's explanation of the actual significance of Clausewitz's conception of absolute war is followed by exposition that demotes the importance of Clausewitz's definition of war as an extension of politics. Although Gallie does not mention Paret by name, he is clearly aiming this discussion at him. Gallie notes that "the accepted view of Clausewitz's philosophy of war is that its core lies in his conception of war as the continuation of policy by the addition of other means, or, more simply, of war as a political instrument." But this view, he argues, "although advocated by genuine admirers of Clausewitz, is liable to mislead," because it suggests that Clausewitz's real interest is not in military questions per se, but in issues "of a wider kind," namely, "politics, and more particularly, in the relations, tensions and struggles between different political units." The reason for such thinking, Gallie maintains, is that it makes Clausewitz "more respectable." "It is widely felt today," Gallie observes, "that, except when it is studied within the wider horizon of politics and sociology, war is a topic as abhorrent in its content as it is weak in theoretical interest" (ibid., 61).

Gallie insists, however, that Clausewitz's primary subject is war as such. "*On War*," he declares, "is emphatically about war, and was primarily written for military men." Clausewitz in *On War*, Gallie is convinced, is not acting primarily as a political theorist. He cites two major reasons for this assessment. In the first place, "Clausewitz's remarks on politics are . . . curiously abstract and meagre." And second, characterizing him as a political theorist would "rob him of his uniqueness and originality," because "others before him had recognized the crucial role of war in politics—Machiavelli, Hobbes, Locke, Montesquieu and Rousseau, to mention only the greatest." Clausewitz, Gallie concedes, does call attention to aspects of war "which are of the first importance for politics, so that *On War* is a work of considerable educative value for practitioners and theorists alike." But "these educative bonuses are always expressed in the most general terms: the last thing Clausewitz would have claimed to be doing in *On War* was to teach politicians their proper business" (ibid., 61–62).

Gallie's analysis does have shortcomings. He fails, for example, to address two issues of fundamental importance. One is Clausewitz's novel concept of augmenting verifiable historical fact with theoretical surmise; another is his contention that the defense is the stronger form of war than the offense. In the case of the relationship between history and theory, Gallie correctly identifies Clausewitz's approach—here surpassing both Aron and Paret in insight—but then he chooses not to explain its significance. Gallie similarly mishandles Clausewitz's views on the superiority of the defense over the attack, which is to say that he touches on the subject perceptively but without the follow-through that would be required to establish its critically important function. Gallie's analytical reticence in this area requires further explanation.

Gallie's assessment of the opening chapter of *On War* is incomplete. As he himself stated, his examination, which is meant to resolve problems arising out of the definitions of absolute and real war, focuses on its opening and conclusion (sections 1–4 and 23–28), with no attention paid to the middle. In effect, this means virtual nonengagement with Clausewitz's statements about defense as the stronger form of war (section 17). Gallie does note Clausewitz's views on the need for weak states to be capable of armed resistance against invasion—which implies that inferior military power can be effective against superior military power when used in the act of defense. He also mentions in passing Clausewitz's recognition of guerrilla war. Above all, Gallie achieves a high level of suggestive insight in remarks that connect defensive fighting by a mobilized citizenry to the realization of absolute war in actuality. Clausewitz, he argues, "attributed the vast differences in the intensity of wars to one main cause: the degree of involvement, usually on the ground of national survival, of either or both of the peoples concerned." And, Gallie adds, "he was thus led to equate wars which 'approach the Absolute point' with what we might call 'wars of embattled democracy'" (ibid., 49, 63, 65).

In the same year that *Philosophers of Peace and War* appeared, Gallie published an essay entitled "Clausewitz Today" in the *European Journal of Sociology*. This article covers much of the same ground as the printed version of the Wiles Lecture, but its exposition, freed from the need to engage the larger themes of an anthology of thinkers on armed conflict, presents Gallie's views on Clausewitz and *On War* in sharper relief. Gallie praises the new translation of *On War* for its readability and accuracy, assessments qualified by minor reservations about specific passages, which are described in a long footnote ("Clausewitz Today," 145–146). He also criticizes Liddell Hart for his "misreadings of (or perhaps sheer failure to read through) *On War*," and notes that Aron has "completely demolished Liddell Hart's thesis" (ibid., 147). Gallie enumerates the virtues of Aron's study, endorsing its findings with enthusiasm, with the exception of his treatment of the first chapter of Book I. Here, he maintains that "Aron . . . seems quite blind to the logical confusions which make the revised Book I Chapter I of *On War* one of the most tragically contorted and opaque and unhelpful pieces of reasoning ever produced by a serious philosopher" (ibid., 152).

Gallie celebrates Paret's monograph as a work of "admirable objectivity and unfailing narrative skill," but confesses "to a slight feeling of disappointment over what this book has not done." This includes failure to "emphasize or advertise the originality of his subject's thought" (ibid., 163–165).[35] In particular, Gallie is highly dissatisfied with Paret's resolution of the apparent conflict between absolute and real war as defined by Clausewitz—the same issue which had so vexed Aron. Paret, in a passage quoted by Gallie (and quoted earlier in the section of this chapter on Paret), had argued that "limited wars might be a modification of the absolute, but need not be, if the purpose for which they were waged was also limited. Violence continued to be the essence, the regulative idea, even of limited wars fought for limited ends; but in such cases the essence did not require its fullest possible expression." Gallie does not regard Paret's argument as wrong so much as he sees it as unclear, offering at best only a partial answer to the questions raised by Clausewitz's terminology:

> I *think* I understand what Paret is trying to say in these tortured
> sentences, and I believe it corresponds to one of the positions to
> which Clausewitz's theorizing brought him. But . . . the position in

question is not the *only* one which Clausewitz struggled to
communicate through his unhappy opposition of "absolute" and
"real" wars. (Ibid., 146; italics in original)

Gallie does not speak to the relationship between history and theory in his
article. But he shows greater appreciation of the significance of Clausewitz's
views on defense than he did in his chapter. He recognizes that Book VI is of
critical importance, observing that its contents helped "to bridge the gap
between the wholly revised Book I Chapter I and other ostensibly more one-
sided discussions of the essence or 'element' of war." He criticizes the summary
of *On War* by Bernard Brodie that appears in the new translation for its per-
functory dismissal of "the most original passages in Book VI . . . without any discussion of their relevance to the total structure of Clausewitz's military
thought." And he explicitly frames Clausewitz's notion of defensive action by an
armed citizenry in terms of guerrilla war. Clausewitz, Gallie argues, believed
that in People's Wars there existed a "tendency . . . to approach the Absolute
Form." And he later notes "Clausewitz's advocacy and analysis of guerrilla formations as a supplement, especially in long defensive operations, to the use of
regular troops" (ibid., 147, 149, 160).

Gallie expresses complicated and somewhat obscure views on historical reenactment and the superiority of the defense over the offense. Although he
more or less recognizes their existence as significant components of Clausewitz's thinking, he does not analyze their character in detail. Nor does he explore
their relationships to other major propositions. They are not, for these reasons,
given their due. Gallie senses the existence of great and even defining arguments
lurking in the later books of *On War*. But he is handicapped by an imperfect
command of the text—as indicated by the serious misreading noted previously;
moreover, blinded, perhaps, by the technical sophistication of his own philosophical analysis of Clausewitz's opening chapter, he does not relate his examination of absolute war to matters of equal if not greater importance. That said,
Gallie knew there was much work to be done, and he had definite views on how
that work should proceed.

In *On War*, Gallie observes in his article, Clausewitz anticipated "many of
the best methods of recent philosophy." The philosophical instruments he had
in mind, he says, occupied "some rather hazy borderland between logical and
linguistic studies." Here he undoubtedly would have included the work of Peirce
and Collingwood, about whom he had written at length, and almost certainly

Ludwig Wittgenstein (1889–1951), a near contemporary at Cambridge and the leading proponent of linguistic philosophy, as well. In any case, Gallie calls for the application of post-Clausewitzian advances in philosophy to the "critical study of the logical foundation of (and flaws in) Clausewitz's conceptual system" (ibid., 144, 150, 167):

> I can think of no other [work] to which a competent student of the
> logical and methodological achievements of the last hundred years
> could devote himself more rewardingly. Of course it would be
> necessary for such a student to appreciate that war is a serious matter,
> and as proper an object for philosophical reflection as art or economy
> or mathematics or the experimental and social sciences. Is it too much
> to hope that mastery of the new logical techniques and competence of
> our century is not incompatible with so obvious and so urgent a
> concern with the struggle and trials of mankind? (Ibid.)

Gallie proposed this investigatory agenda in the hope, if not the belief, that the burst of Clausewitzian scholarship of 1976 was only the beginning of a "burgeoning . . . Clausewitz industry" (ibid., 144). This was not to be the case. The books and articles on Clausewitz that followed over the next two decades were few and far between. No one responded to Gallie's call; nor did anyone offer a persuasive major alternative to the comprehensive view of the theoretical dimension of *On War* produced by Aron (Gat, *History of Military Thought*, 172). Gallie believed that "the best of Clausewitz . . . can be recognized and developed only when the flaws in his conceptual system are exposed and adequately corrected" through the deployment of up-to-date philosophical analytical methods ("Clausewitz Today," 145).

In fact, looking at Clausewitz in the light of philosophical methodologies that emerged after his time and up through the twentieth century does not expose Clausewitzian error or provide remedy, but rather reveals a remarkable degree of congruence on important issues. In the next chapter, we will examine selected works of Peirce, Wittgenstein, and Collingwood and review late twentieth-century work in chaos theory and cognitive science to see how these ideas can shed light on Clausewitz's thought. We will also look at Clausewitz's military and political experiences, which were the main sources of his theoretical creativity.

3

Antecedents and Anticipations

WHILE A STUDENT AT THE INSTITUTE in the Military Sciences for Young Infantry and Cavalry Officers in Berlin from 1801 to 1804, Clausewitz learned about the works of Kant, Machiavelli, Montaigne, Montesquieu, and Voltaire.[1] Following mobilization, war service, and exile from 1805 to 1807, he served as a staff officer and instructor at the General War School for four years, during which time his writing engaged the ideas of Machiavelli as well as of Justus Möser, Johann Heinrich Pestalozzi, and Johann Gottlieb Fichte.[2] After the conclusion of the Napoleonic Wars in 1815, Clausewitz served as a staff officer and military education administrator. His postwar social circle included Hegel, whose work he may have known, and other distinguished intellectuals and artists,[3] and his duties at the General War School brought him into contact with accomplished civilian faculty in the sciences (Paret, *Clausewitz and the State*, 310–311). Clausewitz's exposure to art and literature was undoubtedly broadened and deepened by his relationship to Marie von Brühl, a woman of considerable intellect and culture; he courted her from 1803 and the two married in 1810 (ibid., 103).

Clausewitz possessed the temperament of a scholar. In 1807, he wrote to Marie: "I regard it as completely unimportant whether one is a soldier or not, and if I had enough means to do without any work at all I would gladly retire to the country [and] devote myself to the study of history and of war" (ibid., 436n11). In 1815, Clausewitz observed in a letter to General Augustus Wilhelm Gneisenau (1760–1831) that if he were free to choose, "I would gladly throw myself wholly into the arms of scholarship" (ibid., 245).

Clausewitz's broad and serious intellectual interests, and his academic disposition, have formed the basis of contentions that his thinking on military matters was heavily influenced by the nonmilitary thought of the late eighteenth and early nineteenth centuries. Azar Gat goes so far as to argue that "only a small minority of the principal themes propounded by . . . Clausewitz originated within the military field itself," most having been "extracted from, and set in motion by, the ideas and ideals of new and powerful cultural trends" (*History of*

Military Thought, 142). These were, according to Gat, "conceptions of knowledge and reality, man, art, and history" espoused by the "German Movement," which embraced "the major trends of romanticism, nationalism, and idealism" (ibid., 143–144). Gat concludes that "Clausewitz's real intellectual greatness" is attributable to the fact that he based "a most sophisticated formulation of the theory of war" upon a "highly stimulating intellectual paradigm" that "brought the conception of military theory into line with the forefront of the general theoretical outlook of his time" (ibid., 255). One consequence of this point of view is that Gat does not consider how Clausewitz's substantial war service and intensive study of war might have affected his thinking.

Raymond Aron, in contrast, largely dismisses the influence of nonmilitary thought on Clausewitz (*Clausewitz,* 54, 226–232).[4] He does not, however, attempt to relate Clausewitz's specific knowledge of war to his writing about military affairs. This may be because his highly abstract and incomplete representation of *On War* is difficult to connect to particular historical events. "It seems to me both impossible and fruitless," Aron notes in the preface to his biographical remarks, "to write a political and military account of the wars of the Revolution and Empire," and indeed "equally futile" to address the particular Prussian aspects of the story (ibid., 12). Peter Paret briefly describes the character of Clausewitz's involvement in the great wars of his time, acknowledging that these experiences had "taught him a great deal" (*Clausewitz and the State,* 222–223). He devotes the overwhelming bulk of his study of the origins of Clausewitz's thought, however, to careful consideration of the influence of military and nonmilitary writing. W. B. Gallie recognizes that Clausewitz's war experience "gave him an unusually well-balanced insight into the greatest and most terrible military event of his age," and that "it is difficult to believe that the hardship, shortcomings and disappointments of his career did not serve him well when he set about systematizing his ideas on the general conduct of war."[5] But Gallie lacked the space in his chapter and article to give the matter the treatment it deserved.

There is indeed evidence that the military history of Prussia during the Napoleonic Wars informed Clausewitz's belief that the defense is the stronger form of war; moreover, his personal experiences during this period could well have served as the basis of his conception of historical reenactment of the supreme commander under difficult circumstances. The Prussian defeat and recovery in the wars against France, and Clausewitz's analysis of these events, are telling. Following an analysis of this issue, in this chapter I shall compare the conclusions that Clausewitz drew from his experience to similar findings of later philosophers, and then consider Clausewitz's approach to the study of strategic decision-making in light of recent work in mathematics and cognitive science.

Historical Analysis

From 1792 until 1815, Europe was torn by a succession of major wars. France's military power, enhanced by the political and social effects of the revolution of 1789, and magnified by the extraordinary leadership of Napoleon Bonaparte (1769–1821), was at the root of these conflicts. The major military histories of the Napoleonic phase of the period have focused upon the exploits of this great soldier. As a consequence, historians have usually framed the story of fighting in early nineteenth-century Europe in terms of the rise and fall of French predominance. Clausewitz's perspective was different. His main concern was the catastrophic defeat and national subjugation of Prussia at the hands of France, and its subsequent revival and liberation through military reform and the support of allies. Clausewitz's historical studies of Prussia's major campaigns emerged from these developments.[6]

For several years, Europe enjoyed peace in spite of the revolution of 1789 in France. International tranquility ended, however, when the French National Assembly declared war against Austria in April 1792. That summer, Prussia joined Austria in an invasion of France. Prussian hopes of swift victory and easy territorial gains were dashed, however, by the unexpectedly strong stand of the French army at Valmy in September. Unwilling to assert its full strength against France, and with its appetite for expansion satisfied in part by the partitions of Poland, Prussia made peace in 1795. Prussia's former continental allies soon followed suit, leaving England to carry on the fighting alone. Prussia refused to join the second coalition against France, which went to war in 1798 and ended with a general peace in 1802. Prussia at first declined to participate in a third coalition, but secretly changed its position after the outbreak of hostilities in 1805. Before Prussia could act openly, however, Napoleon's brilliant victories over the Austrian army at Ulm and over the combined forces of Austria and Russia at Austerlitz forced Austria to seek terms.

The defeat of Austria left Prussia, whose hostile intentions were known to Napoleon, exposed to French attack. France enjoyed two great military advantages—its army heavily outnumbered that of Prussia and, blooded by recent victorious hard action, was more experienced. Russia, for its part, had withdrawn its army to Poland. Napoleon was convinced that Prussia's sovereign,

Frederick William III (1770–1840) would, given the circumstances, agree to large concessions in order to avoid war. This proved to be correct. In February 1806, the Prussian king reluctantly ratified the Treaty of Schönbrunn, which forfeited territory, allied Prussia to France, removed Prussia's anti-French chief minister, and closed Prussian ports to British commerce. In compensation, Prussia was given Hanover, a large, populous, and prosperous state that was the ancestral property of the king of England. In June 1806, however, Napoleon offered to return Hanover to Britain in exchange for a peace settlement. This diplomatic overture not only failed to achieve its main objective but precipitated a drastic change in Prussian policy.

By mid-1806, several developments had promoted the growth of a strong anti-French faction within the Prussian court. These included Napoleon's humiliating treatment of Prussia, increasing French involvement in German affairs, and the war with Britain, which had come about because of Prussia's agreements with France and had serious consequences for Prussian commercial prosperity. Centered around Queen Louise (1776–1810), the anti-French faction included a number of influential senior army officers. The king and others in his entourage were inclined toward continued accommodation with France. They suspected, however, that France's ultimate objective was nothing less than the destruction of Prussia. These fears were confirmed by news of Napoleon's negotiations with Britain over Hanover. In early August 1806, Frederick William III decided to go to war and ordered the mobilization of the army. In early September, Prussian forces entered Saxony, which convinced Napoleon, who had previously been uncertain of Prussian intentions, that hostilities were imminent. The swift mobilization of the French army followed. In late September, Prussia issued a formal ultimatum, with a reply demanded by 8 October. France responded with invasion.

The Prussian military deployment was slow and indecisive. There were several reasons for this. The king had neither the training nor the experience to command an army. The army high command lacked recent war experience, and the senior generals were elderly and divided in opinion. Moreover, the three officers sharing the office of chief of staff differed in their views of what needed to be done. It is thus hardly surprising that Prussian councils of war often resulted in protracted and confused discussion. A tendency to disfavor defensive courses of action meant that proponents of withdrawal eastward to unite with Russian reinforcements, or a lesser withdrawal northward to a position that flanked the expected line of any French advance, were unable to carry their views. The decision to launch a preemptive offensive, however, presumed that France would require considerable time to prepare an attack, and would thus initially assume a defensive posture. When this assessment proved erroneous, the lack of a general staff, and the inadequate staff organization at the lower corps and divisional levels, delayed and disrupted the issuing of orders. As a consequence, the reorientation of the Prussian army to meet the French attack was both tardy and in important respects misdirected.

On 8 October, the French army entered Saxony, moving northeast directly toward Berlin. Such movement threatened the Prussian capital and would put French forces between the main Prussian army, which had been deployed to move west toward the Rhineland, and Russian forces moving west from Poland. The Prussian high command thus abandoned its planned offensive and in effect adopted a course it had previously rejected, placing their forces on the flank of the French line of advance. During the hasty redeployment, Prussia's Saxon ally suffered a minor defeat. This occurred at the battle of Schleiz on 9 October. An isolated Prussian division, inadvertently placed in the path of the French army, was shattered the next day at the battle of Saalfeld. The French juggernaut continued its forward movement. When its left wing reported the presence of major Prussian forces around Erfurt, Napoleon reacted by shifting the direction of movement from northeast to northwest in order to engage and destroy the Prussians as soon as possible.

On 13 October, a single corps, commanded by Napoleon, encountered the Prussian rearguard under the Prince of Hohenlohe (1746–1818) at Jena. Within twenty-four hours, the French had been reinforced by elements of the Imperial Guard, additional cavalry formations, and three additional corps. The rapidity of the French concentration, which Hohenlohe did not anticipate, meant that his force was compelled to fight on 14 October at a heavy numerical disadvantage. By the time his belated request for reinforcements had been answered by the arrival of a reserve contingent, the battle was lost. Napoleon believed he had defeated the Prussian main body. But in fact, this task was accomplished by a single French corps under the command of Marshal Louis Nicholas Davout (1770–1823), which had been ordered northward to cut off what Napoleon believed would be the Prussian line of retreat. Instead, Davout's corps caught the Prussian main army under the Duke of Brunswick (1735–1806) on the march near Auerstädt on 14 October. Although heavily outnumbered, Davout beat off poorly coordinated Prussian assaults, during which Brunswick was killed. King Frederick William III assumed command, but his incompetent direction contributed to further Prussian setbacks. When Davout's vigorous counterattack threatened both Prussian flanks, the king ordered a retreat.

Prussian losses in both engagements were enormous. Of the 40,000 troops deployed at Jena, 10,000 were killed or wounded and 15,000 made prisoner. Of the 63,000 men fielded at Auerstädt, 10,000 died and 3,000 were captured. Losses of equipment were no less disastrous. Of the 298 pieces of artillery the Prussian army had started with, 235 fell into the hands of the French during the course of these two battles. In other words, in one day of fighting Prussia lost nearly 40 percent of its engaged manpower and more than three-fourths of its committed artillery. Although Davout's corps suffered heavy casualties, Napoleon's main body did not. Moreover, one French corps, under Marshal Jean-Baptiste Jules Bernadotte (1763–1844), had not been involved in either battle and was thus able to spearhead a rapid and vigorous pursuit of the retreating Prussians. Over the course of the next thirty days, the French army took an additional 122,000 prisoners, captured 1,800 more guns, and occupied Berlin. By the second week of November, the Prussian army had been all but destroyed.

Prussia's subsequent military effort was feeble. The only significant formation left was a weak corps that managed to join forces with the advancing Russians in December. Although it performed well at the hard-fought battle of Eylau in February 1807, this was not enough to prevent a Russian retreat. Nor was the Prussian remnant strong enough to hold Königsberg after the Russian defeat at the battle of Friedland in June 1807. The Prussians were excluded from the peace negotiations between Napoleon and the Russian tsar, Alexander I (1777–1825), at Tilsit that began on 25 June. Peace discussions between France and Prussia were not concluded until after the ratification of the Franco-Russian agreement on 7 July, and the separate treaty was not ratified until 12 July. The terms imposed were no less humiliating than the diplomatic process that produced them: Prussia lost half its territory and population, and French troops occupied much of what remained.

Prussia's catastrophic defeat prompted fundamental military change. In July 1807, Frederick William III formed the Military Reorganization Commission, headed by General Gerhard Johann David von Scharnhorst (1755–1813). Scharnhorst had favored a defensive strategy in 1806 and was the Prussian army's leading proponent of reform. In August 1808, the king accepted the recommendations of the commission to open the officer corps to nonaristocrats and to moderate army discipline in order to promote popular support for military service. The king also agreed in principle to the commission's call for universal conscription. Fiscal exigency, however, barred large expansion of the army, and at the insistence of the French, the Paris Convention of September 1808 formally limited the Prussian army to less than a quarter of its prewar size and prohibited the formation of a militia. In the spring of 1809, Scharnhorst became chief of the general staff. From this position he directed the improvement of the Prussian army's organization, tactical doctrine, officer education, and troop training.

The fiscal and treaty limitations on Prussian army expansion caused the Prussian leadership to consider seriously how to counterbalance French military superiority. In the summer of 1808, Spanish and British forces successfully challenged French control of the peninsula, in part owing to the contributions of armed civilians fighting in Spain. Inspired by the Spanish example, Augustus Wilhelm Gneisenau, a member of the Military Reorganization Commission, proposed that an armed population be encouraged to fight in the event of renewed hostilities with France. The king, however, rejected such a course because he believed it would promote political activity incompatible with the maintenance of royal and aristocratic authority, thus undermining general social order. In the spring of 1809, however, the concept of state-sponsored guerrilla warfare gained adherents when it seemed as if Prussia might support Austria's war with France. Prussia in the event remained neutral. But in 1811, French opposition to Prussian military reform, and a large increase in the number of French troops in Germany, suggested that hostilities between Prussia and France were imminent. These circumstances, in combination with the continued relative weakness of the Prussian regular army, strengthened the case for a defensive alliance with Russia, whose relations with France had deteriorated sharply since Tilsit, and the preparation of the general population for armed resistance to French occupation.

The debate on both courses of action was rendered moot by a dramatic reversal of Prussian policy. The preservation of what little was left of Prussian national sovereignty required either hostilities with France, which would unquestionably involve heavy losses and risk total defeat and complete subjugation, or cooperation with France, which, while distasteful, offered the immediate prospect of survival. Frederick William III chose the latter course. On 24 February 1812, the Treaty of Paris committed Prussia to participate in a French invasion of Russia. When Napoleon launched his assault on Russia in June 1812, his army of 655,000 included a contingent of 20,000 Prussians. Napoleon hoped to achieve a decisive victory within twenty days by destroying the main Russian forces on the frontier. Instead, ill-coordinated movement of widely separated French forces and misjudgment of enemy intentions allowed the Russian armies to escape, avoiding major engagement. In August, the French won the battle of Smolensk, but the Russian army retreated in good order. In September, French victory at the battle of Borodino enabled Napoleon to occupy Moscow, the Russian capital, but the Russian army remained intact and Tsar Alexander refused to come to terms.

The French position was untenable. Napoleon's main army had lost nearly two-thirds of its strength during the campaign thus far. By October it was outnumbered by Russia's main army, which had been reinforced after Borodino. The French forces disposed to protect his flanks, which included the Prussian contingent, were at an even greater numerical disadvantage. In mid-October, after concluding that the Russians were determined to fight on, Napoleon belatedly ordered a general withdrawal. The Russians pursued, inflicting losses in harassing attacks, while fatigue, sickness, and starvation caused further heavy casualties. In mid-November, the onset of winter magnified French difficulties enormously. Although the Russian army proved incapable of blocking Napoleon's retreat, French losses to hit-and-run attacks by Russian cavalry—and the effects of cold—were very great. By the end of the campaign in January 1813, 370,000 of the original invading force of 655,000 men had died, with most of the remainder either wounded or taken prisoner.

In December 1812, the Prussian contingent under the command of General Johann David Ludwig Yorck (1759–1830) was serving as a rearguard to the retreat of the French forces that had protected the northern flank of Napoleon's main army. Threatened by envelopment, Yorck was receptive to entreaties by Prussian officers serving in the Russian army to come to terms. Without consulting his king, he agreed to the Convention of Tauroggen on 30 December 1812, which in effect withdrew his army from the war. Yorck's action inspired proponents of resistance to France, who by this time had widespread support in Prussia. King Frederick William III reluctantly abandoned his policy of cooperation with Napoleon. On 28 February 1813, the secret Convention of Kalisch allied Prussia with Russia, and on 13 March, Prussia declared war on France. Prior to this, several factors had helped to lay the foundation for rapid expansion of the army. These included clandestine efforts to build up a reserve of trained soldiers, increases sanctioned by the French to provide additional troops for service against Russia, and the adoption of conscription. By the spring of 1813, Prussia could field 135,000 men.

Chapter Three

The Prussian army was commanded by Marshal Gebhard Leberecht von Blücher (1742–1819), with Scharnhorst as his chief of staff. Together with the Russian army, it fought two major battles against the French in May 1813. In the first, at Lützen, General Scharnhorst was mortally wounded; the other was at Bautzen. Although the French won both engagements, the reformed Prussian army fought well, and both the Prussian and Russian armies withdrew in good order. A truce mediated by the Austrians suspended hostilities in early June. By the end of the cease-fire in August, the Prussian army had grown to over a quarter of a million men, which was larger than it had been in 1806. With the resumption of the war, moreover, Prussia and Russia were joined by Austria and Sweden. Over the course of the 1813 campaign, Napoleon weakened his army by making several ill-advised and unsuccessful attempts to capture Berlin, the capital of Prussia. These actions set the stage for his defeat at the battle of Leipzig in October 1813 at the hands of the combined armies of the four powers previously named, which resulted in the dissolution of French control of Germany. In addition, the heavy losses in men and materiel sustained by the French in 1813 compounded the effects of the losses they had suffered in Russia and Spain. This exposed France to allied invasion, which began in December. In February and March 1814, Napoleon won a series of brilliant tactical victories against poorly coordinated Prussian and Russian armies. In spite of these setbacks, Napoleon's opponents persevered. As a consequence, French strength gave out, resistance crumbled, and the defensive front collapsed. In March 1814 the allies occupied Paris. In April, Napoleon abdicated power.

From May 1814 until February 1815, Napoleon lived in exile on Elba, an island off the west coast of Italy. Prompted by news of discontent in France, he returned to France with a thousand men of his personal guard. By late March he had entered Paris and taken control of the government. Britain, Austria, Prussia, and Russia responded with an alliance and the mobilization of their armies. Napoleon could have stood on the defensive, using time to build up his army, which would have enabled him to meet the coalition invasion with much stronger forces than had been available to him the previous year. Instead, he chose to strike offensively with the smaller army in hand in the hopes of disrupting the alliance by defeating the armies of Britain and Prussia in the Netherlands, which were as yet unsupported by those of Russia and Austria. The Anglo-Dutch army was under the command of the Duke of Wellington (1769–1852). Von Blücher commanded the Prussian army, with Gneisenau as his chief of staff.

In June 1815, the swift advance of Napoleon's army caught the British and Prussians by surprise. The armies of Wellington and von Blücher were too far apart to offer mutual support, and the Prussians had not been able to concentrate all of their forces. On 16 June, two French corps engaged elements of the Anglo-Dutch army at Quatre Bras, the latter retreating to Waterloo. On the same day, the main body of the French army under Napoleon took on the Prussian army at Ligny, inflicting disproportionately heavy losses and compelling it to withdraw. On 18 June, Wellington's army fought a defensive battle against the main French army under Napoleon, who had detached two corps under Marshal Marquis Emmanuel de Grouchy (1766–1847) to pursue the Prussian army. At Wavre, a Prussian corps under the command of General Johann Adolf von Thielmann (1765–1824)—acting as a rearguard for the main Prussian army, which was moving to support the Anglo-Dutch army at Waterloo—fought a successful delaying action against Grouchy's superior forces. Blücher's troops arrived at Waterloo in time to win the battle and executed a vigorous pursuit, which destroyed the French army. Napoleon's second abdication and final exile followed in July 1815.

Clausewitz wrote detailed military histories of much of the period just described, most of which were produced in the years immediately preceding his composition of *On War*. This body of work was based upon his participation in the campaigns in question as well as on personal communication with Prussian and Russian senior commanders during and after the fighting and published accounts. Between 1823 and 1825 he composed *Observations on Prussia in Her Great Catastrophe,* a critical analysis of the events leading up to and including the disastrous campaign of 1806.[7] During this same period, he assembled the text of *The Campaign of 1812 in Russia* from material he had drafted between 1814 and 1823. In the summer of 1813, at the request of Gneisenau, he had written a patriotic history of the 1813 campaign in Germany. In the early 1820s he also wrote two monographs on the 1814 campaign, *Strategic Critique of the Campaign of 1814 in France* and the shorter *Summary of the Campaign of 1814 in France.* Finally, in 1827 he wrote *The Campaign of 1815 in France.* Prussia did not resort to national insurrection by armed civilians; this was nonetheless a potential course of action between 1808 and 1812. Clausewitz was a major proponent of guerrilla warfare, and his views on the subject during this period were recorded.[8]

Clausewitz was probably predisposed to write histories of Prussia's struggle with France in a way that illustrated the propositions that defense was a stronger form of war than offense, and that this was especially so when the state could count upon the political and therefore military support of the general population. As a student at the Institute for Young Officers, he had almost certainly read the work of Count Friedrich Wilhelm Ernst zu Schaumburg-Lippe-Bückeburg (1724–1777), who had been an important teacher of Scharnhorst, who in turn was Clausewitz's teacher and patron. The count favored the defense over the offense on moral grounds, and both he and Scharnhorst believed that small states could defeat larger ones through strategic defense based upon the adoption of mass conscription.[9] Clausewitz also demonstrated an early appreciation of the fact that traversing great distances could incapacitate an attacking army. In 1804 he observed that "if Bonaparte should someday reach Poland he would be easier to defeat than in Italy, and in Russia I would consider his destruction as certain."[10]

At first glance it may appear that it would be difficult, if not impossible, to derive a clear lesson about the relative merits of the defense and the offense from the history of the Napoleonic Wars from 1806 to 1815. From the French perspective, vigorous offensive action had triumphed in 1806 and then failed disastrously in 1812. From the Prussian point of view, offensive action had misfired in 1806 and succeeded in 1814 and 1815. Clausewitzian historical analysis, however, did not focus exclusively on events as they occurred, but took into account the circumstances that conditioned action and the significance of alternative courses. This approach enabled Clausewitz to draw favorable conclusions about defensive action even in cases of successful offensives, and to characterize successful offensives as the counteroffensive phase of a defense. When these considerations are taken into account, along with Clausewitz's notion of the strategic significance of guerrilla warfare, what emerges is a picture of the historical foundations of the proposition that "the defense is the stronger form of war."

In 1806, Scharnhorst was the leading proponent of a defensive strategy that would have involved withdrawal eastward in order to combine the armies of Prussia and Russia before taking on the French invasion. Scharnhorst was a patron and friend of Clausewitz,[11] but at this time Clausewitz appears to have favored energetic offensive action. He not only welcomed war as an opportunity to win distinction, but just before Auerstädt he drafted a plan for an immediate counteroffensive.[12] His enthusiasm for strategic attack evaporated after the disastrous outcome of the 1806 campaign and its consequences, which weakened

Prussia to the point that an offensive strategy was out of the question. In 1808, he argued that in the event of a renewal of hostilities against France, Prussia should avoid decisive battle against superior forces, even if this meant the forfeiture of all national territory and the retreat of the army onto foreign soil.[13] This attitude informed his historical analysis of the 1806 campaign. In his *Observations of Prussia in Her Great Catastrophe*, Clausewitz concludes that Prussia's best course in 1806 would have been to withdraw the army as Scharnhorst had wished.[14]

In his account of the French invasion of Russia in 1812, Clausewitz argues that the Russian victory was produced by two separate dynamics. The first is the disproportionate losses suffered by the attacker while advancing. Clausewitz observes that during its retreat, the Russian army fell back slowly on its lines of supply, which minimized casualties caused by fatigue or hunger, and that reinforcements more than made up for the casualties that did occur. In contrast, the forward movement of the French army extended its lines of supply, and the difficulty of resupply on the march was exacerbated for the French by the fact that the Russians had destroyed magazines and crops along the way. The French alleviated the problem to a degree by shifting a large portion of the army onto side roads in order to widen the zone of foraging, but movement along paths unsuitable for heavy traffic was slow and tiring (*Campaign of 1812*, 175–180). Although French combat casualties were considerable, the large reduction in French fighting strength during their advance, Clausewitz shows, was mostly attributable to privation, and fatigue, and the need to garrison occupied territory.[15] His concludes that

> the French army reached Moscow already too much weakened for the attainment of the end of its enterprise. For the facts that one third of its force had been wasted before reaching Smolensko, and another before Moscow, could not fail to make an impression on the Russian officers in command, the Emperor, and the ministry, which put an end to all notion of peace and concession. (Ibid., 100)

The second dynamic is the exploitation of French overextension by robust counterattack. Clausewitz is convinced that the virtual destruction of the French army during its retreat was primarily due to vigorous pursuit by the Russian army. Although he recognizes that cold, hunger, and fatigue reduced fighting efficiency for the French and even resulted in a considerable number of fatalities, he insists that it was armed action that converted injury into annihilation. The "actions at Wiazma, Krasnoi, and the Beresina," Clausewitz

90

Chapter Three

concludes, " . . . occasioned enormous losses to the French; . . . whatever critics may say of particular moments of the transaction, the entire destruction of the French army is to be ascribed to the unheard-of energy of the pursuit, the results of which imagination could hardly exaggerate" (ibid.).[16] Clausewitz recognizes that the French disaster could have been even greater. The Russians came close to completely destroying the French main body and capturing Napoleon at the Beresina River crossing in late November 1812. "Never," Clausewitz observes of this event, "were circumstances more propitious towards reducing an army to capitulate in the field." If the Russian generals had not been so reluctant to try conclusions with an intact—albeit weakened—force headed by a man they regarded as a military genius, Napoleon may not have escaped; as it was, the war was prolonged for two more years (ibid., 210–212).

The successful Russian response to the French invasion, Clausewitz argues, was a matter of circumstances rather than by design. The Russians were initially forced to withdraw because of overwhelming French superiority on the frontier (ibid., 14). In the early stages of the campaign, Clausewitz attempted to convince General Karl Ludwig von Phull, the tsar's chief military adviser, that retreating deep into the interior would overextend the French army and expose it to a crushing riposte, but Phull was not persuaded (ibid., 27–28). From time to time, Russian generals sought to stand and fight, hoping to stem the French tide, but they either suffered defeat, as at Smolensk, or retreated due to unfavorable conditions (ibid., 42, 106, 114). Withdrawal, in other words, "arose out of the mere pressure of circumstances" rather than from "any distinct consciousness on the part of the Russian chiefs" (ibid., 131). Once the French army was critically weakened, which was readily apparent by the fall, counterattack in force was an obvious course. At this stage, Clausewitz argues, "things began of themselves to work for the Russians, and a good result was inevitable without much exertion on their part" (ibid., 141–142).

Clausewitz concedes that the correctness of retreat and counterattack in the Russian case could only have been understood beforehand "by a man of extended views, clear understanding, and rare greatness of mind" (ibid., 141). Even early in the campaign, however, he had been highly critical of Russian attempts to justify taking offensive action against the French before they attacked, in the belief that such a course would be premature. Russian attack at that point, he was convinced, would have involved great risks incommensurate with the possible gains. Clausewitz notes that the Russians believed that "great advantage belonged to the offensive" and that the "Russian soldier was more adapted for attack than defense." He dismisses these notions, ironically observing, "It is known that all armies assert this of themselves." "Whoever investigates the subject," Clausewitz subsequently declares, "will say to himself that the offensive form is the weaker, and the defensive the stronger, in war; but that the results of the first, when successful, are positive, therefore the greater and more decisive; of the latter only negative, by which the equilibrium is restored, and one may be advocated as well as the other." That the two might be combined is perhaps implied but not stated (ibid., 114–120).

Clausewitz recognizes that Russia's great size favored a strategy of retreat followed by counterattack:

> The Russian realm is so large that we may play at hide and seek in it with an enemy's army, and this fact must be the groundwork of its defense against a superior enemy. A retreat into the interior draws an enemy after it, but leaves so much territory behind him that he cannot occupy it. There is scarcely a difficulty then for the retreating army to retrace its march towards the frontier, and to reach it *pari passu* with the then weakened force of the enemy. (Ibid., 147)[17]

The story of the Russian campaign of 1812, Clausewitz believes, had general implications. An attempt to take over even a small country could pose problems of occupation that could be sufficient to frustrate the designs of a much stronger attacker. The Russian example, he writes, had confirmed an existing conviction that "it is impossible to obtain possession of a great country with European civilization otherwise than by aid of internal division" (ibid., 184).

The proposition that a smaller country could frustrate conquest even when faced by a militarily much superior opponent is closely related to Clausewitz's views on the strategic significance of action by scattered forces. After the catastrophic defeat in 1806, Prussia's leading civilian and military reformers made the concept of "People's War"—defined as state-orchestrated fighting waged by soldiers and armed civilians in territory occupied by a foreign power—a central feature of their strategic thought.[18] Faith in the efficacy of such a method was in large part inspired by what was believed to be the effectiveness of Portuguese and especially Spanish resistance to French invasion in the Peninsular War, which had broken out in 1807. Interest in and knowledge of the war in southwestern Europe was fueled by firsthand accounts of former Prussian officers engaged in the fighting—including Scharnhorst's son.[19] In the spring of 1808, widespread action by bands of soldiers and civilians operating behind French lines facilitated, in what came to be known as "guerrilla war," the victories of the Spanish regular army and a British expeditionary force in Portugal. Guerrilla warfare thereafter played a significant role in a protracted and bitter contest that weakened French military power.[20]

At first, Clausewitz harbored reservations about the desirability, if not the efficacy, of guerrilla warfare as an instrument of state power.[21] By 1810, his position on the subject seems to have become more favorable. From October 1810 through June 1811, he taught a course at the General War School on *kleiner Krieg* ("little war"). He formally defined his subject as "the use of small troop units in the field." "Actions involving 20, 50, 100, or 300 or 400 men make up the little war," he said, "unless they are forming part of a wider battle."[22] For the most part, his exposition was concerned with the use of regular troops. Although Scharnhorst, then army chief of staff, was probably sympathetic to consideration of guerrilla warfare,[23] Clausewitz did not overtly discuss action by armed civilians at this point. This is probably because he felt it necessary to avoid the displeasure of the crown and the disapproval of certain conservative military superiors, who would have been opposed to the inclusion of talk on guerrilla warfare in the formal instruction of junior officers.[24] Such prudence may also have dictated caution in his instructions to the Crown Prince of 1812. In that communication, he did not mention guerrilla warfare directly, but he perhaps was alluding to its efficacy when he wrote that "in defensive warfare even the means of small states are infinitely great."[25]

In his private correspondence, Clausewitz was less guarded. In 1811, when war between an all-powerful France and a partially occupied and militarily weakened Prussia seemed imminent, he advised Gneisenau that his task as a commanding general would be to make "a Spain out of Silesia."[26] In February 1812, Prussia agreed to support a French war against Russia. Clausewitz was adamantly opposed to such a course,[27] which prompted him to abandon caution and, among other things, describe his views on guerrilla warfare openly. In his memorandum to the king, the Bekenntnisdenkschrift ["Political Declaration"] of February 1812,[28] Clausewitz calls not only for national conscription, but for action by an armed citizenry as well. The French, Clausewitz declares, should expect "a Spanish civil war in Germany" ("Bekenntnisdenkschrift," I: 729).[29]

Resorting to guerrilla war, Clausewitz recognized, could provoke strong French countermeasures, as it had in Spain, but he did not believe that harsh French action would break the will to resist. Loss of civilian courage, he argues, was "an unnecessary worry!" He continued:

As if we could not be just as cruel as the enemy and as if the enemy
was not made from the same flesh and blood as us! The enemy will try
this method and the war will quickly take on a cruel nature. But to
whose disadvantage? Obviously to the disadvantage of the side that
cannot risk heavier losses because they are fighting with regular forces.
Let us dare to repay cruelty with cruelty and to revenge violent acts
with violent acts. It will be easy . . . to bid higher than the enemy and
thus force him back within the boundaries of restraint and humanity.
(Ibid., I: 733–734)

Clausewitz justifies the use of "People's War" on the grounds that the counterinsurgency effort would greatly reduce the number of French forces available to
contend with Prussian regular troops and those of any ally, thus paving the way
for decisive victory (ibid., I: 732–734).

Clausewitz's conception of state-sponsored resistance to occupation by
bands of soldiers and armed civilians was never actualized, as the conflict
between Prussia and France was forestalled by the alliance against Russia in 1812.
The next year, Prussia was able to free itself with regular and militia forces aided
by the armies of powerful allies. In his account of the 1813 campaign, Clausewitz
describes the national spirit that would have been required to animate a general
revolt against a foreign occupation, namely, "Inveterate and renewed hatred im-
bibed against the tyrant and oppressor of legal liberty and civic freedom."[30]
Clausewitz also notes that a French defensive strategy in April 1813 would almost
certainly have defeated an allied attack, and that the French army was critically
weakened by subsequent allied defensive actions in the late spring.[31] The polemical requirements of Clausewitz's account of the 1813 campaign, a work intended
to inspire patriotic enthusiasm, precluded criticism of the Prussian leadership.
But after the battle of Bautzen, Clausewitz, who was then attached to the staff of
the Prussian army, had vehemently called for an immediate Russo-Prussian
counterattack, a course that was rejected.[32] Had Clausewitz written an objective
account of the 1813 campaign on the lines of his other histories, there can be little doubt that he would have said that the failure to reap the fruits of a successful
defense through a follow-up offensive had constituted a major strategic error.[33]

In his study of the allied invasion of France, Clausewitz concludes that Napoleon ultimately lacked the strength "for the kind of defense waged by offensive means" that he had practiced with success in the early stages of the 1814
campaign,[34] but that his performance nevertheless illustrated, in Paret's words,

"the potential strength of the defensive."[35] In his analysis of the Waterloo campaign of 1815, Clausewitz argues that in principle, a defensive strategy offered Napoleon the best hope of victory. Such a course, he observes, would have gained Napoleon time to mobilize further national strength, weakened the attackers by forcing them to garrison occupied territory, and allowed patriotic fervor to provoke a national uprising against the invaders. In actuality, however, Napoleon did not enjoy the full support of the French people, whose political division practically ruled out the possibility of mounting a guerrilla war as a productive auxiliary to action by regular forces. Under these circumstances, according to Clausewitz, Napoleon had no choice but to take the offensive.[36]

Clausewitz applied standard military analytical techniques to the study of well-known facts of recent history in order to formulate major propositions about the relative strengths of the defense and offense and the strategic potential of combining the two. Though his conclusions contradicted the propensity among most officers of his day to favor the attack over the defense, his objective methodology was essentially conventional, and his argument thus easy to comprehend, if not accept. The situation was practically the reverse with respect to Clausewitz's approach to the dynamics of high command in a major war. Clausewitz's contention that strategic decision-making by the commander of an army involves complex, contingent, and difficult problem-solving is acceptable on its face. His manner of dealing with this question, however, requires him to reject conventional military history and objective military analysis and replace them with a subjective form of historical reenactment. Why and how Clausewitz conceived of such an expedient is the subject of the next section.

Philosophical Invention

As in the case of his views on defense as the stronger form of war, Clausewitz's approach to the dynamics of high command was probably shaped by his early education. Here, again, the influence of Scharnhorst was critical, if not defining. Scharnhorst was convinced that the survival of the Prussian state in the face of the threat posed by Napoleonic France required a supreme commander of extraordinary ability. He furthermore believed that the pool of candidates from which such a leader would be chosen could be enhanced significantly through military education. And Scharnhorst also insisted, according to Paret, that "in the absence of experience . . . the study of war had to be based on history." Scharnhorst's views on the importance of history, which were a manifestation of his distrust of pure theory, were especially significant. "No military theorist of the time," Paret observes, "was as conscious as Scharnhorst of the innate conflict between theory and reality. His elaboration of this fundamental issue, and his refusal to seek its solution in increasingly complex abstractions, constitute the most important lesson he taught Clausewitz" (Paret, *Clausewitz and the State*, 67–68, 71).

The effect of such instruction is apparent in Clausewitz's earliest major writing. In his study of the Swedish campaigns in north Germany during the Thirty Years' War, for example, composed between 1803 and 1806, Clausewitz criticizes the general practice of writing military history that does not take adequate account, in Paret's words, of the "psychology of the commander, his ambitions, his awareness of his own abilities." These qualities, Clausewitz believes, were "the most decisive." He therefore focuses his strategic analysis on the personality of Gustavus Adolphus, the Swedish commander-in-chief, and the psychology of his Catholic opponents. "The manner of 'Gustavus Adolphus,'" Paret observes, "forecast Clausewitz's subsequent historical method and style." Clausewitz also emphasizes the importance of the psychology of the commanding general in his essays on strategy of 1804. He insists, moreover, that the best method of learning generalship is the proper study of history. Clausewitz, Paret concludes, believed that such intellectual work is necessary to enable "men of independent and powerful character" to "liberate themselves from dogma," "suit their actions to their special circumstances," and "willingly resort to battle," a quality which, "potentially or in reality, is at the basis of all wars" (ibid., 85, 87, 90).

From 1810 to 1812, Clausewitz was military tutor to Crown Prince Frederick William (1795–1861). In his written instructions to the prince, he expressed his views on the critical importance of moral factors in no uncertain terms. Theory, he declares, not only took into account physical factors, such as the preponderance of army strength or the advantages of position, but also moral powers, such as "the probable errors of the enemy, upon the impression made by a bold spirit of enterprise, &c.&c.—even upon our own desperation."[37] Clausewitz further advises his charge that "some great sentiment must stimulate great abilities in the General," mentioning ambition, hatred, and pride. "Open your heart," he writes, "to a feeling of this kind." Further, he says, "Be bold and astute in your designs, firm and persevering in executing them, determined to find a glorious end, and destiny will press on your youthful brow a radiant crown—fit emblem of a Prince, the rays of which will carry your image into the bosom of your latest descendents.[38]

During the wars against Napoleon, Clausewitz's combat experience at the tactical and operational levels was considerable. At the battle of Auerstädt, his battalion was involved in heavy fighting until, confronted by an overwhelming force of the enemy, it was compelled to surrender (Parkinson, *Clausewitz*, 61–63, 68–76). During the initial Russian retreat from their western frontier in the summer of 1812, he acted briefly as chief of staff to Count Peter Pahlen, whose forces constituted the rearguard of the Russian army (ibid., 147–151). He subsequently was assigned to the First Cavalry Corps of Count Feodor Petrovich Uvarov (1773–1824) as quartermaster general, and he served with this unit at the battle of Borodino (ibid., 161). While on the staff of Prince Ludwig Wittgenstein (1769–1843), Clausewitz participated in the winter pursuit of the French army, witnessing the infamous mass slaughter of wounded and noncombatants at the Beresina River in November 1812 (ibid., 193–194). He was present at the battles of Lützen and Bautzen in May 1813; during the former he led two cavalry charges and was involved in a desperate fight when surrounded by French infantry (ibid., 219–220). As chief of staff to Count Ludwig Georg von Wallmoden, commander of a Russian army corps that included a powerful German contingent, Clausewitz participated in a number of minor actions in northern Germany in 1813 and 1814.[39] And in 1815, as General von Thielmann's chief of staff, Clausewitz was on the field at Ligny and played a major role in the direction of the hard-fought rearguard action against Grouchy at Wavre.[40]

Many of Clausewitz's generation could claim comparable records. What makes Clausewitz unique is the fact that throughout most of this period, he was placed in positions from which he could either observe strategic decision-making firsthand or learn about the nature of strategic decision-making from those who had exercised high command. As Scharnhorst's main assistant from 1808 to 1810 (and unofficially to 1811), he was privy to the memories and thoughts of the chief of staff of the army and de facto minister of war, and thus to the inner workings of military strategy during a critical period. In 1812, Clausewitz was attached to the Russian general staff as aide-de-camp to Major General Karl von Phull, a Prussian officer upon whom Tsar Alexander relied for strategic advice. In December 1812, Clausewitz was an eyewitness of—and indeed a key actor in—General von Yorck's decision at Tauroggen to neutralize the Prussian forces attached to the French army. In 1813, Clausewitz became a liaison officer for the Russian army on General von Blücher's staff, a role that allowed him to act as Scharnhorst's assistant until his death after Lützen. He then briefly advised Gneisenau, who replaced Scharnhorst as Blücher's chief of staff, before being reassigned to a secondary front. In the fall of 1815, Clausewitz became chief of staff to Gneisenau, who commanded the newly constituted Army of the Rhine, and even after Gneisenau's brief tenure maintained a regular correspondence with his former chief.[41]

In addition, Clausewitz's understanding of the interrelationship between strategy and politics must have been influenced by his personal contacts with the highest levels of the national leadership in Prussia and Russia during the wars against France. From 1803 to 1809, he was aide-de-camp to Prince August, a cousin of the King of Prussia. This placed him at the center of Prussian military and political affairs, which enabled him to observe or learn about the character of individuals, and not just follow the course of events. Clausewitz's connections to the court were further enhanced by preferment and his private life. In October 1810, as mentioned previously, he was appointed military tutor to the Crown Prince. In December 1810, Clausewitz married Marie von Brühl, whose high social standing and intelligence meant she was capable of reporting to her husband on political activity at court. In the spring of 1813, Clausewitz is known to have had intense discussions about strategy with King Frederick William III after the battle of Bautzen. While in the service of the Russian army in 1812, Clausewitz discussed strategy with the tsar on more than one occasion.[42]

Clausewitz could have drawn upon his remarkable experiences to write a lively "insider" memoir or, like Jomini, used them to formulate a complete system of doctrine. He probably never considered the former course, which would have been beneath his dignity, and he had been schooled by Scharnhorst to disdain the utility of the latter. This probably explains Clausewitz's difficulties when he came to write a work of theory that identified the major elements of strategy. Clausewitz produced a draft of such a work between 1816 and 1818 while he served with the Army of the Rhine at Coblenz. In a note describing the history of this manuscript, Clausewitz states that his original intention had been to write "concise, aphoristic chapters" that "would attract the intelligent reader by what they suggested as much as by what they expressed." But he then admits that his tendency "to develop and systematize . . . completely ran away" with him, which caused him to elaborate "as much as I could." He also reveals that he planned to revise his text in order to "strengthen the causal connections in the earlier essays, perhaps in the later ones draw together several analyses into a single conclusion, and thus produce a reasonable whole."[43]

Clausewitz's intentions were extraordinarily ambitious. "I wanted at all costs to avoid every commonplace," he declares, "everything obvious that has been stated a hundred times and is generally believed. It was my ambition to write a book that would not be forgotten after two or three years, and that possibly might be picked up more than once by those who are interested in the subject."[44] He was apparently dissatisfied with this effort, however. No text of the manuscript has survived except for the preface. But it would not be unreasonable to surmise that the attempt to systematize and develop disparate concepts to produce a single conclusion had resulted in what amounted to a statement of doctrine, or something that was too close to it for comfort. At this point in his intellectual development, Clausewitz apparently had yet to solve the problem posed by the inadequacy of concepts, and indeed of language itself, to deal with the moral—that is to say, psychological—requirements of strategic decision-making.

In the preface, Clausewitz raises the issue of the superiority of direct observation over theoretical description. "It would obviously be a mistake," he writes, "to determine the form of an ear of wheat by analyzing the chemical elements of its kernel, since all one needs to do is to go to a wheat field to see the grown ears." He hoped that, by basing his axioms "on the secure foundation either of experience or the nature of war as such," he would succeed in carrying out the latter approach when it came to understanding war. But Clausewitz seems to have thought his effort was unsatisfactory, even though it avoided the worst features of existing theories. Attempts by previous writers "to make their systems coherent and complete," he states, had resulted in works "stuffed with commonplaces, truisms, and nonsense of every kind." To illustrate his complaint, Clausewitz provides the text of a satirical piece lampooning pretentious exposition that does nothing more than state the obvious.[45] But as for his own work, he warns the reader that his book does not contain "a complete theory," but "offers only material for one."[46]

After a decade of further study and reflection, Clausewitz was able to develop a work that addressed the moral requirements of making difficult decisions in a manner that avoided platitudinous theorizing. This is *On War*. In the course of presenting his ideas, he makes clear his dissatisfaction with the way in which language was commonly used to convey meaning about the nature of strategy. Strategy, he observes, is about action, not simply "statements" or "declarations" as in "other areas of life." Clausewitz then says that "words, being cheap, are the most common means of creating false impressions" (*On War* [Howard/Paret], III/10: 202). He criticizes existing military theory for its "retinue of *jargon, technicalities, and metaphors*," which "swarm everywhere—a lawless rabble of camp followers."

Clausewitz believes that the proper use of language required simplicity and awareness of its limitations, promising to "avoid using an arcane and obscure language, and express ourselves in plain speech" (ibid., II/5: 168; italics in original). Clausewitz later writes:

> The reader expects to hear of strategic theory, of lines and angles, and instead of these denizens of the scientific world he finds himself encountering only creatures of everyday life. But the author cannot bring himself to be in the slightest degree more scientific than he considers his subject to warrant—strange as this attitude may appear. (Ibid., III/7: 193)

Clausewitz was convinced that definitions of terms could be no more than approximations of the real phenomena they were supposed to represent. "The nature of the question," he maintains, "makes it impossible to give an accurate definition of space, mass, and time." But "the fact that these concepts cannot be more accurately defined should not be considered a disadvantage," he says, because "unlike scientific or philosophical definitions, they are not basic to any rules" (ibid., V/2: 280–281). "Such expressions," Clausewitz observes,

> as "a dominating area," "a covering position," and "key to the country" are, insofar as they refer to the nature of higher or lower ground, for the most part hollow shells lacking any sound core. These elegant elements of theory have been used above all as seasoning for the apparently overly plain military fare. They are the favorite topics of academic soldiers and the magic wands of armchair strategists. (Ibid., V/18: 354)

"We want to reiterate emphatically," he warns, "that here, as elsewhere, our definitions are aimed only at the centers of certain concepts; we neither wish nor can give them sharp outlines. The nature of the matter should make this obvious enough" (ibid., VI/27: 486).

In *On War,* Clausewitz problematizes the utility of conventional approaches to conceptualization as well as the usefulness of specialized vocabulary. "Theory," he observes, "becomes infinitely more difficult as soon as it touches the realm of moral values." The difficulty, Clausewitz is convinced, is that the emotional condition of an individual—a factor that he believed is critical with respect to decision-making under difficult circumstances—cannot be accurately reduced to a concept that is universally valid. "Moral values," he observes, "can only be perceived by the inner eye, which differs in each person, and is often different in the same person at different times" (ibid., II/2: 136, 137). Moral quantities "will not yield to academic wisdom. They cannot be classified or counted. They have to be seen or felt" (ibid., III/3: 184). "We might list the most important moral phenomena in war," Clausewitz writes,

> and, like a diligent professor, try to evaluate them one by one. This method, however, all too easily leads to platitudes, while the genuine spirit of inquiry soon evaporates, and unwittingly we find ourselves proclaiming what everybody already knows. (Ibid., III/3: 185)

Given the incommunicable nature of an individual's moral condition through words or ideas, *practical* instruction in the formulation and implementation of strategy cannot be delivered directly. "A book," Clausewitz laments, "cannot really teach us how to do anything" (ibid., II/3: 148).

To surmount the limitations of language and ideas with respect to the accurate description of the moral dynamics of an individual, Clausewitz invented an indirect approach to understanding decision-making in war. It considers both the physical and moral factors that conditioned decision-making in particular historical cases—including those which must be surmised—in order to evoke a sense of command dilemma and the moral forces required to act decisively in spite of it. Because meaning cannot be explained directly, it has to be conjured through a form of psychological reenactment. Clausewitz calls such an exercise "critical analysis," but the bare phrase does not do justice to its essential character. Exercises in critical analysis can generate different feelings and thus even different solutions in different individuals or even the same individual at different times. The critical variable, in other words, is the mental state of the person doing the reenactment, and the critical outcome is the development of the individual's comprehension of the nature of high command in its moral as well as its physical aspects. Reenactment, in other words, is a form of personal psychological experiment. Experiencing moral dilemma and surmounting it emotionally as well as intellectually may have been what Clausewitz means when he uses the word "scientific." As he had observed in 1818, "there is no need today to labor the point that a scientific approach does not consist solely, or even mainly, in a complete system and a comprehensive doctrine."[47]

On War does not, therefore, teach its reader how to acquire a skill as such, but rather how to explore realms of personal thought that included emotional elements in relation to the sorts of difficult problem-solving likely to arise in the course of decision-making in war. The intended product of this process is a sensibility—that is, a mental character encompassing emotion as well as knowledge—that, in the absence of actual experience, can provide a measure of sound understanding and a platform for further learning. Clausewitz's method of learning thus involved a process of inducing a form of self-knowledge, as opposed to the importation of technical knowledge. It was prompted not only by the recognition that a scientific approach might consist of something other than the formulation of laws of cause and effect, but also by the understanding that language is incapable of accurately representing the phenomenon of high command, that language can nonetheless be used as a medium of productive intellectual work, and that the specific character of this work should take the form of psychological reenactment of past events. To a remarkable degree, these propositions preconfigure important aspects of the thought of later major philosophers.

W. B. Gallie, as explained previously, noted the resemblance of Clausewitz's philosophical methods to those developed by others after his death. Though Gallie does not refer to Charles Sanders Peirce by name, Gallie's own monograph on Peirce and, as will be demonstrated, the affinities in the thought of Peirce and Clausewitz, make it practically certain that Gallie has Peirce in mind when he writes that Clausewitz's work would have been "appreciated by some of the ablest philosophers of our century."[48] It is likely that this category includes Ludwig Wittgenstein as well, to whom Gallie is probably referring when he writes of "some rather hazy borderland between logical and linguistic studies"[49] and reports that Clausewitz had anticipated "many of the best methods of recent philosophy."[50] Gallie explicitly states that Clausewitz "anticipates R. G. Collingwood's idea of history as the re-enactment of past deeds."[51] We shall examine each of these important thinkers in turn.

Charles Sanders Peirce (1839–1914) was born in Boston, the second of five children (four sons and a daughter) of Benjamin Peirce (1809–1880), a professor of astronomy and mathematics at Harvard University. Charles Peirce exhibited brilliant ability in a wide range of intellectual activity, including philosophy, experimental psychology, and several of the physical sciences. He was, moreover, an accomplished dramatic reader and actor. Although his physical ailments and character flaws caused self-destructive behavior that impeded and ultimately destroyed his scientific and scholarly career, leading American contemporaries, such as his close friend William James and his student Josiah Royce, recognized his genius, and he was highly regarded in Europe as well. Max H. Fisch, an authority on Peirce's work, called him "the most original and the most versatile intellect that the Americas have so far produced."[52]

Peirce, however, never formulated a unified statement of his ideas. By the end of his life he had published roughly 12,000 pages in the form of papers, articles, and reviews. Surviving manuscripts come to 80,000 pages, and the first attempt to produce a substantial selection of Peirce's published and unpublished writing filled eight volumes. A more complete and better-organized edition currently underway will consist of thirty volumes.[53] Although much has been written about Peirce, the enormous quantity of material, the lack of a well-ordered, comprehensive edition of his writing, and the complexity and difficulty of his thought all contribute to the fact that the identification and critical evaluation of his ideas is still a work in progress. The main burden of representing relevant aspects of Peirce's writing has therefore rested upon Gallie's monograph *Peirce and Pragmatism,* first published in 1952 and republished in a revised edition in 1966. This book provides a reasonably compact and therefore convenient evaluation of important aspects of Peirce's thought, and its findings undoubtedly formed the basis of any comparisons between Clausewitz and Peirce that its author may have made.

The concluding chapter of Gallie's book addresses Peirce's ideas about what constitutes sound scientific method. Peirce, according to Gallie, aims "at totally reorienting our ordinary conceptions" of, among other things, "the nature of general laws and their relation to the things or events which conform to them." Peirce notes that, on one hand, happenings might occur in such a regular way as to indicate propensities or patterns, which he calls "habits." On the other hand, he contends, "'chance spontaneity' affords a sufficient explanation of certain actually observable results." Though Peirce believes that over time the play of habit will increase and that of chance decline, he also thinks this will not occur completely until some "infinitely distant future." Thus all laws are subject to exception to greater or lesser degrees in the form of "arbitrary sporting," "objective chance," or "pure spontaneity" (Gallie, *Peirce and Pragmatism,* 216–222).[54]

This being the case, it would be wrong to declare a body of theory predictive in any absolute sense, he says, although such a thing could usefully "unify or 'thicken' our conceptions of different strands of cosmic or terrestrial or biological or human history." The validity of the predictive approach, Peirce believes, assumes a subject whose dynamics are fixed and for which there can be only a single explanation. The nonpredictive approach, in contrast, posits a subject whose character is potentially the product of any one of a multitude of dynamics determined by highly variable particular circumstances, whose explanation will thus also be highly variable and particular. Theory does not exist in this instance to predict, but to facilitate discovery and observation of particular dynamics as the prerequisite to understanding. Such nonpredictive theory is "genuinely scientific in spirit" because it fosters investigation, while predictive theory is not, because it has the effect, "sooner or later, of 'blocking the road of Inquiry.'" Peirce's concept of what is scientific and why, Gallie notes, is his "central thought in all his cosmological writings" (ibid., 238–240).

Peirce, like Clausewitz, is deeply concerned with the way in which language and ideas can or cannot represent reality. This issue, indeed, is fundamental to his conception of Pragmatism, the philosophical school of which he is widely considered to have been the founder. "Consider," he writes in 1878, "what effects, that might conceivably have practical bearings, we conceive the object of our conception to have. Then, our conception of these effects is the whole of our conception of the object." Those who called themselves Pragmatists understood the phrase, which they believed defined their creed, "to emphasize (though with varying intentions) the importance of practical considerations, of actual or possible physical operations, in even the most strictly intellectual pursuits" (ibid., 11–12). Gallie argues that Peirce sees it quite differently, however:

> Peirce is maintaining that an abstract formula has meaning if, and only if, we can use it,—or can act under its influence in a distinctive and appropriate way. Whether anything—an idea—corresponding to the formula can be found among the furniture of our mind is, on his view, altogether irrelevant to the question whether that formula means anything. Our answer to this question depends solely upon how we answer the prior questions: How does one use the formula, or what distinctive things must one do in employing it, or interpreting it correctly? (Ibid., 15)

Peirce believes, in Gallie's words, that "the meaning of any word, sentence, or other symbol essentially requires a succession of experiences—actions, expectations, and adjustments, real or imagined—to articulate it." Or, to put it even more simply, "the distinctive meaning of a word or other symbol cannot be contained in or equated with a single mental image or picture" (ibid.).

Context, in short, is everything, and this being the case, language is highly susceptible to misuse. Meanings that are productive of communication in one setting can be utterly misleading in another even though the user of the words is unaware that this is the case. This might not matter a great deal in everyday conversation about the commonplace, but it can compromise the integrity of scholarly discourse. In explanation of Peirce's thought, Gallie writes:

> Mathematicians and metaphysicians alike are liable, just because of the highly abstract character of the questions they pursue, to fall into highly sophisticated confusions, with the result that in certain contexts they use expressions which in fact lack the distinctive meaning which they would claim for them. And exactly the same thing is liable to happen in the more theoretical development of the so-called "empirical" sciences: physics, chemistry, biology, psychology—all have ghosts or skeletons in their cupboards, phrases or formulae which long seemed to possess genuine empirical meanings, but which in fact contributed in no way to the established truths of any of these sciences. (Ibid., 142)

Peirce devotes a great deal of attention to the problem of formulating appropriate antidotes to linguistic malpractice, clarification of which would require lengthy exploration of technical and difficult areas of his thought, much of which has little bearing on Clausewitz (ibid., 142–180). Enough has been said to demonstrate the affinity, if not congruence, of the views of Clausewitz and Peirce with respect to sound scientific method and the problem of language use, and thus lend credence to Gallie's suggestive remarks that a significant resemblance existed. The question of how Clausewitz's solution to the problem posed by language as a conveyor of meaning in the case of important matters pertaining to war anticipated later serious philosophical work is better addressed by examining R. G. Collingwood's notion of historical reenactment. But it is worth first exploring the later work of Ludwig Wittgenstein, whose thoughts about language in many ways paralleled those of Peirce, and whose conclusions about how this affected the teaching of activity that could be called performance is relevant to Clausewitz's thinking about this subject.

Ludwig Josef Johann Wittgenstein (1889–1951) was the last of eight children (five sons and three daughters) of Austria's leading steel magnate. He was, like his mother and siblings, a gifted musician, and in general he possessed a highly developed artistic sensibility. Wittgenstein's secondary schooling was technical, and his initial choice at the university was to study aeronautical engineering. Wittgenstein abandoned his engineering studies in 1912 after developing an interest in philosophy, a subject he pursued at Cambridge University, and within two years he had established his reputation as a philosophical genius. During World War I, Wittgenstein served in the Austrian army with distinction, winning several decorations for outstanding bravery. After several years of teaching elementary school, for which he was temperamentally unsuited, and subsequent unemployment, Wittgenstein returned to Cambridge in 1929, where he was awarded a doctorate for previously published work. Wittgenstein was appointed university lecturer and then a fellow of Trinity College in 1930. In 1939 he was made professor of philosophy, a post he held until his retirement in 1947. His collected writings, lecture notes, and correspondence fill more than twenty volumes. Wittgenstein is widely regarded as the preeminent philosopher of the twentieth century.[55]

Wittgenstein's main concern in his late work is the problem posed by the incapacity of language to convey meaning accurately in many important cases. The resemblance of his subject and approach to aspects of Peirce's work may not have been entirely coincidental. Peirce had been an important influence on Frank P. Ramsey,[56] a close friend and intellectual confidant of Wittgenstein. Ramsey was critical of Wittgenstein's early philosophical writing in ways that may have contributed a good deal to his consideration of the language problem in his late work.[57] In any case, Wittgenstein, like Peirce, argues that the meaning of individual words is not intrinsic, but depends upon their use in sentences, and that their meaning even in the context of sentences is in turn heavily dependent upon nonlinguistic nuance. Wittgenstein regards complex emotions, among other things, as one of the most important elements of nuance. Words and sentences are therefore not just conveyors of intended and discrete meaning, but emotional events. Accurate perception of emotional content, and thus of the meaning of even well-chosen words, however, requires extraordinary knowledge of human nature. Reciprocally, inappropriate use of words can obscure, misrepresent, or even obliterate emotional content essential to the accurate conveyance of meaning. And, of course, there is the problem posed by the need to discuss matters that are ineffable with inadequate instruments of expression.

Wittgenstein believes that accurate perception of emotional content cannot be systematized. His thoughts on this matter are worth quoting at length, not only for what they have to say about the importance of wisdom as the prerequisite to accurate perception of human emotion, but also as an indicator of Wittgenstein's attitude toward the viability of what Peirce would have called "predictive theory" and the limitations of language as a medium of accurate communication. Wittgenstein writes,

> Is there such a thing as "expert judgment" about the genuineness of expressions of feeling?—Even here, there are those whose judgment is "better" and those whose judgment is "worse."
>
> Corrector prognoses will generally issue from the judgments of those with better knowledge of mankind.
>
> Can one learn this knowledge? Yes; some can. Not, however, by taking a course in it, but through *"experience."*—Can someone else be a man's teacher in this? Certainly. From time to time he gives him the right *tip.*—This is what "learning" and "teaching" are like here.—What one acquires here is not a technique; one learns correct judgments. There are also rules, but they do not form a system, and only experienced people can apply them right. Unlike calculating rules.
>
> What is most difficult here is to put this indefiniteness, correctly and unfalsified, into words.[58]

For Wittgenstein, teaching certain subjects with words is difficult to the point of being practically impossible. Indirect learning by means of language is one thing, direct learning from experience another. Wittgenstein believes that the nature and meaning of music, for example, cannot be defined or explained adequately in words, but can be demonstrated by the performance of music.[59] In the case of an abstract idea, where demonstration is impossible, reflection on personal experience can induce comprehension where it previously did not exist:

> Life can educate one to a belief in God. And experiences too are what bring this about; but I don't mean visions and other forms of experience which show us the "existence of this being", but, e.g., sufferings of various sorts. These neither show us an object, nor do they give rise to conjectures about him. Experiences, thoughts,—life can force this concept on us.[60]

But the notion that experience can serve as a preceptor where explanation in words cannot raises the question of whether language can be used to induce something resembling experience. Which brings the present discussion to Collingwood.

Robin George Collingwood (1889–1943) was the only son of poor but cultured parents. His father was an archaeologist and artist who became the professor of fine art at University College, Reading, and his mother was a musician. Collingwood inherited his interests and talents from both. He was a gifted pianist, violinist, and composer; an accomplished painter; and an expert archeologist of Roman Britain. At Oxford University, he proved equally adept at both philosophy and history, and because of his outstanding academic performance he was made a fellow of Pembroke College in 1912. During World War I he served in naval intelligence. After the war, he returned to academic life, and in 1935 he became the Waynflete Professor of Metaphysical Philosophy at Oxford. Ill health stemming from intellectual overexertion, which appears to have begun in his undergraduate years, undermined his work. In spite of debilitation, Collingwood wrote prolifically and profoundly on philosophical and historical subjects. From 1938, however, he was crippled by strokes, which forced him to retire in 1941. Collingwood died in 1943.[61]

A man of broad erudition, Collingwood appears to have read at least part of *On War,* but there is insufficient evidence to determine the extent to which his thought was thereby influenced.[62] The resemblance of Collingwood's thoughts about language to those of Wittgenstein have been noted,[63] but little appears to be known about any communication between the two, other than the fact that Collingwood disapproved of Wittgenstein's work for reasons that are unknown.[64] What distinguishes Collingwood's approach to language from that of Peirce and Wittgenstein is the manner in which he integrates his concerns about the limitations of language with what he believes constitutes a representation of the past that is superior to conventional historical narrative. Language deployed carefully and with due recognition of its limitations, Collingwood believes, can be used to reenact the mental processes of historical figures, and such reenactment constitutes the only kind of history that can offer sound instruction in practical matters.

Collingwood holds that ordinary language is the best medium of communication about things that are not strictly technical, such as those that are the subject of historical or philosophical inquiry. "The business of language," he writes in his *Essay on Philosophical Method* (1933),

> is to express or explain; if language cannot explain itself, nothing else can explain it; and a technical term in so far as it calls for explanation, is to that extent not language but something else which resembles language in being significant, but differs from it in not being expressive or self-explanatory. (*Philosophical Method,* 207)[65]

A writer about things philosophical, Collingwood argues, thus has to avoid technical language and adopt terminology that possesses "that expressiveness, that flexibility, that dependence upon context, which are the hall-marks of a literary use of words as opposed to a technical use of symbols" (ibid., 207). The attitude of a philosopher, Collingwood knows, is essential when attempting to understand the behavior of human beings in the past, which constitutes much of what is called history, because of certain shortcomings in the common forms of historical discourse.

He maintains that historians preserve narrative cohesion and the authority of the narrator's voice at the cost of introducing serious distortions in the presentation of the past as a human experience. He observes that they

> try to steer clear of doubts and problems, and stick to what is certain. This division of what we know into what we know for certain and what we know in a doubtful or problematic way, the first being narrated and the second suppressed, gives every historical writer an air of knowing more than he says, and addressing himself to a reader who knows less than he. . . . The writer, however conscientiously he cites authorities, never lays bare the processes of thought which have led him to his conclusions. (Ibid., 209–210)

Philosophers, in contrast, write about those things omitted by the historian. He notes that the philosopher's purpose

> is not to select from among his thoughts those of which he is certain and to express those, but the very opposite: to fasten upon the difficulties and obscurities in which he finds himself involved, and try, if not to solve or remove them, at least to understand them better. . . . The philosopher therefore, in the course of his business, must always be confessing his difficulties, whereas the historian is always to some extent concealing them. (Ibid., 210)

Whether a text is identified as essentially philosophical or historical is a matter with serious implications for the reader, Collingwood says:

> In reading the philosophers, we "follow" them: that is, we understand what they think, and reconstruct in ourselves, so far as we can, the processes by which they have come to think it. There is an intimacy in the latter relation which can never exist in the former. What we demand of the historian is a product of his thought; what we demand of the philosopher is his thought itself. (Ibid., 211)

Thus a philosophical work is not a form of prose, but rather what Collingwood calls "a poem of the intellect." What this poem expresses is "not emotions, desires, feelings, as such, but those which a thinking mind experiences in its search for knowledge; and it expresses these only because the experience of them is an integral part of the search, and that search is thought itself" (ibid., 212). "The reader, on his side," Collingwood goes on to state later on,

> must approach his philosophical author precisely as if he were a poet, in the sense that he must seek in his work the expression of an individual experience, something which the writer has actually lived through, and something which the reader must live through in his turn by entering into the writer's mind with his own. (Ibid., 215)

Collingwood's insistence upon the critical importance of experience, and the experience of inquiry into experience, as the source and measure of any presentation of human affairs, including those of the distant past, inform his writing on the philosophy of history. In an essay on the nature and aims of the philosophy of history (1924–1925), he rejects in principle the notion that history can be used to construct general laws and specifically rules out the legitimacy of doing so with regard to the conduct of warfare. Collingwood also maintains that knowledge of past events is always incomplete in significant ways, observing, of the Battle of Hastings, that "no one knows, no one ever has known, and no one ever will know what exactly it was that happened."[66] But in 1928, according to his autobiography, Collingwood formulated his concept of reenactment, which in place of exact historical knowledge—which is impossible to obtain—offers a form of productive verisimilitude about past events that he is convinced could justifiably be called scientific. In his words,

> I expressed this new conception of history in the phrase "all history is the history of thought." You are thinking historically, I meant, when you say about anything, "I see what the person who made this (wrote this, used this, designed this, &c.) was thinking." Until you can say that, you may be trying to think historically but you are not succeeding. And there is nothing else except thought that can be the object of historical knowledge. Political history is the history of political thought: not "political theory," but the thought which occupies the mind of a man engaged in political work: the formation of a policy, the planning of means to execute it, the attempt to carry it into effect, the discovery that others are hostile to it, the devising of ways to overcome their hostility, and so forth. . . . Military history . . .

is not a description of weary marches in heat or cold, or the thrills and chills of battle or the long agony of wounded men. It is a description of plans and counter-plans: of thinking about strategy and thinking about tactics, and in the last resort of what the men in the ranks thought about battle. (*Autobiography*, 110)[67]

Reenactment as a historical method is based on three conditions. First, the original "thought must be expressed" in either language or some other activity that provides an indication of the historical actor's state of mind. Second, "historical knowledge is the re-enactment in the historian's mind of the thought whose history he is studying." Here, Collingwood makes a distinction between the actual thought of the historical figure and the reenacted thought of the historian, which are not identical. Reenacted thought is, in Collingwood's parlance, "encapsulated"—that is, thought with awareness that it is about a past event and not about present concerns. And third, "historical knowledge is the re-enactment of a past thought encapsulated in a context of present thoughts which, by contradicting it, confine it to a plane different from theirs." Here, Collingwood specifies the nature of the difference between the original and encapsulated thought: The former is directed toward the solution of a problem in the past, and the latter involves the replication of past problem-solving processes in order to improve problem-solving thought in the present. "We study history," Collingwood thus declares, "in order to see more clearly into the situation in which we are called upon to act" (ibid., 111–114).

At the heart of Collingwood's conception of reenactment is the alteration of the thought processes of the reenactor through the replication of past thought:

If what the historian knows is past thoughts, and if he knows them by re-thinking them himself, it follows that the knowledge he achieves by historical inquiry is not knowledge of his situation as opposed to knowledge of himself, it is a knowledge of his situation which is at the same time knowledge of himself. In re-thinking what somebody else thought, he thinks it himself. In knowing that somebody else thought it, he knows that he himself is able to think it. And finding out what he is able to do is finding out what kind of a man he is. If he is able to understand, by re-thinking them, the thoughts of a great many different kinds of people, it follows that he must be a great many kinds of man. He must be, in fact, a microcosm of all the history he can know. Thus his own self-knowledge is at the same time his knowledge of the world of human affairs. (Ibid., 115)

In coming to this conclusion, Collingwood believes he has answered an important question prompted by World War I, which he regarded as a man-made catastrophe that could have been avoided had statesmen thought otherwise than they had. "How," Collingwood had asked, "could we construct a science of human affairs, so to call it, from which men could learn to deal with human situations as skillfully as natural science had taught them to deal with situations in the world of Nature?" "The answer," he writes, "is clear and certain. The science of human affairs was history" (ibid.). In saying this, he is thinking of history as a kind of worldly wisdom rather than as a narrative account of events. To the question, "Of what can there be historical knowledge[?]," Collingwood, in his classic study *The Idea of History*, answers "that which can be re-enacted in the historian's mind," which "must be experience."[68]

These brief glimpses into the thought of Peirce, Wittgenstein, and Collingwood provide ample evidence for Gallie's contention that Clausewitz anticipated important later philosophical work and possessed original philosophical talent of a very high order. Like Clausewitz, all three thinkers problematized language with respect to the communication of meaning about matters involving human behavior, distrusted the invention of technical vocabularies, were skeptical of the utility of theory that was based upon rules, and believed that experience can convey meaning in ways that language cannot. Like Clausewitz, Peirce and Collingwood explicitly addressed the matter of what constitutes a scientific approach to the study of human affairs. And although full consideration of Clausewitz's conception of historical reenactment is reserved for the next chapter, enough has been said thus far to show that it strongly resembles Collingwood's thinking in both form and function.

Clausewitz was both a man of thought and a man of action. Although he devoted himself to writing about military matters, he also had a strong appreciation for the arts. Peirce, Wittgenstein, and Collingwood also had wide interests that balanced action with thought and exhibited robust artistic inclinations. In writing about the nature of things, these men may have entertained similar views simply because of the manner in which they engaged the world. They had something else in common as well: All four lived through wars considered to be the greatest conflict in their nation's history. Although Peirce exhibited indifference[69] — which does not preclude his having been affected — Wittgenstein and Collingwood, like Clausewitz, were participants, and Wittgenstein, like Clausewitz, was a soldier in combat. Thus the congruences in their philosophical outlook may in some indefinable but nevertheless significant way have owed something to the fact that they were all polymaths with artistic sensibilities who lived their formative adult lives in the shadow of catastrophic war.

Clausewitz's concern with the psychology of supreme command made him a student of human nature and the public affairs of men under the special conditions of serious fighting. The interests of Peirce, Wittgenstein, and Collingwood were more general but essentially the same, as they focused on the nature of man as an actor in a social universe. But just as examining the work of these three philosophers can help to put Clausewitz's arguments on the moral dimension of war into perspective, consideration of twentieth-century science can facilitate understanding of his thought about the dynamics of strategic decision-making. In recent years, chaos theory has offered mathematical explanations of complex phenomena such as war, while the study of the workings of the mind through the integration of biology and psychology has provided a productive framework for the study of how people make choices about courses of action under difficult circumstances. In the next section we shall therefore discuss salient work in these areas as a means of further illuminating Clausewitz's approach to the exercise of supreme command.

Scientific Perspective

In "Clausewitz, Nonlinearity, and the Unpredictability of War," an article published in 1992,[70] Alan Beyerchen, a professor of German history at Ohio State University, related advanced mathematics to the study of Clausewitz. In the article he argues that Clausewitz viewed war as a "nonlinear" phenomenon, maintaining that much of the difficulty in understanding Clausewitzian argument can be attributed to the fact that this perspective is at variance with that of all other major theorists of war and with that of most readers of military theory, who view war as a linear phenomenon. He concludes that in order to improve comprehension of *On War*, readers need to recognize this difference of perspective with respect to what constitutes sound theory, delegitimize what has been the old normative view, and legitimize the Clausewitzian approach by making it the new normative view. The reasoning that supports Beyerchen's important propositions deserves fuller explanation.

In mathematics, Beyerchen observes, "linear" describes "a system of equations whose variables can be plotted against each other as a straight line." In such a system, outputs are proportional to inputs—that is, "small causes produce small effects" and "large causes generate large effects." In addition, "the whole is equal to sum of its parts"—that is, a problem can "be broken up into smaller pieces that, once solved, can be added back together to obtain the solution to the original problem." "Nonlinear" describes a system that exhibits "erratic behavior through disproportionately large or disproportionately small outputs," or involves interactions that result in the whole not being equal to the sum of the parts. The problem of nonlinear systems, however, is that their behavior over time cannot be described by equations that are easy or even possible to solve. For this reason, analysts of many different kinds of nonlinear phenomena have idealized nonlinear systems as linear in order to allow the deployment of mathematics capable of producing a solution. In such cases, Beyerchen observes, "reality has been selectively addressed in order to manipulate it with the tools available" ("Nonlinearity," 62, 64).

Beyerchen concedes that idealized linearization works reasonably well for nonlinear systems that are stable. But he rejects the viability of this approach for nonlinear systems that are not stable. Stability is defined in terms of the susceptibility of the condition of the system to change. If very small forces can cause much larger effects, the system can be regarded as unstable. When a system is said to be "sensitive to initial conditions"—and thus unstable—this means that "immeasurably small differences in input can produce entirely different outcomes for the system, yielding various behavior routes to a degree of complexity that exhibits characteristics of randomness." The source of randomness, high levels of which are called "chaos," is not dependent upon the number of variables and their general nature, but upon the possibility of wide variation in the particular values of the variables and the resulting multiplication to infinity of possible different interactions, qualities of interaction, and outcomes. One could thus describe a chaotic system's general dynamics accurately, which could be theoretically useful, yet be unable to predict its behavior (ibid., 65, 66).

Beyerchen argues that Clausewitz's three definitions of war are all "prominently marked by nonlinearity" and presume unstable conditions. In the first case, that war is "an act of force to compel our enemy to do our will," nonlinearity is a function of the complex interactions that arise out of the contest of opposed military actions. In the second case, that war is "merely the continuation of policy by other means," nonlinearity is generated by the complexity and dynamism of interactions among military circumstances that are changing as well as by concomitant alterations in political objectives that in turn can change military behavior. In the third case, that war is a "remarkable trinity" consisting of emotion, chance, and rationality, nonlinearity is again the product of complex and changing interactions among the three variables (ibid., 66–72). Although a certain aspect of Beyerchen's discussion of the third definition is problematical, about which more later, his general conclusions with respect to all three definitions are convincing.

Having demonstrated that Clausewitz views war as a nonlinear and unstable phenomenon, Beyerchen explains how such a characterization is embodied in the main exposition of *On War*. The manifestation of nonlinearity in this work, Beyerchen argues, is a matter of his "emphasis on unpredictability." Clausewitz, he says, addresses three primary sources of unpredictability. First, Clausewitz throughout *On War* is concerned with unpredictability arising out of interactions "between adversaries" and "in processes that occur on each side as a consequence of the contest." Second, Clausewitz argues that unpredictability is generated by what he calls "friction," ultimately deriving from the behavior of the individuals constituting the army, whose capacity for conflict with each other or even with the commander-in-chief is bound to reduce the efficiency of an army's capacity to act to a greater or lesser degree. And third, Clausewitz is aware that in war unpredictability is frequently the product of chance, and thus, that "in the whole range of human activities, war most closely resembles a game of cards" (ibid., 72–79).

Beyerchen states emphatically that although Clausewitz considers war to be essentially nonlinear, he nonetheless recognizes the existence of forms of war that can be described in linear terms. "As much as any military professional," Beyerchen observes, "[Clausewitz] clearly wants to find or generate conditions under which outcomes may be guaranteed." For Clausewitz, such exceptions to the norm defined by nonlinearity occur when political considerations sharply constrain the use of violence. Clausewitz also embeds "linearity in a general environment of nonlinearity" when he argues that the effects of victory will be in proportion to the scale of victory, but that this relationship of cause and effect can be altered by certain factors, such as "the character of the victorious commander" or the possibility that defeat will invigorate rather than demoralize the vanquished (ibid., 82–84).

To readers accustomed to thinking of theory in linear terms, Clausewitz's juxtaposition of nonlinear and linear characterizations of war seems to create a "maddening maze of qualification." Most readers, Beyerchen argues, labor "under the implicit imperative that a good theory must conform to a linear intuition" and, in the face of such apparent confusion, choose to perceive *On War* in simplified and therefore linearized terms. In doing so, however, they fail to engage Clausewitz's characterization of war as essentially nonlinear and never come to appreciate the concomitant fact that Clausewitzian theory in *On War* would be valid only if this were the case. As a consequence, the essential character of Clausewitzian theory is completely misunderstood. Moreover, the nature of the misunderstanding allows readers to sustain their faulty intuitive position that all theory must be linear (ibid., 84–87).

Beyerchen draws four conclusions from this line of reasoning. First, "full comprehension of the work of Clausewitz demands that we retrain our intuition." This means that we must reject linear analysis as the basis of military theory. Second, this rejection will require "a reevaluation of Clausewitz as an authority in military manuals and training." Instead of using declarative phrases from *On War* as statements of principle, more attention will have to be paid to "Clausewitz's forest of caveats and qualifications," which "more faithfully represents the conditions and contexts we actually encounter." Third, we must pursue a more complex understanding of the relationship between war and politics. In place of the widespread notion that the role of politics is to set the objectives of a war at the outset, with these same objectives dictating the course of the subsequent fighting, politics should be seen as a variable that is subject to continuous change as it affects and is affected by military acts. And finally, chance is not a factor that influences war, but is intrinsic to war. War is not a system affected by chance, but a system whose interactive dynamics generate chance, and because this is the case, any theory that accurately represents this characteristic cannot be predictive (ibid., 88–90).

Beyerchen's characterization of Clausewitzian theory as nonlinear is very convincing and also useful. By identifying the nonlinear argumentation of *On War* and demonstrating its incompatibility with the linear mindsets of most readers, Beyerchen provides an explanation for why this work has been so misunderstood. And misunderstanding Clausewitz's work is tantamount to misunderstanding armed conflict itself, since his nonlinear approach offers a better explanation of the nature of war than linear theory does. Beyerchen's findings have serious implications for military affairs and military education. "Theorists," he writes, "must not be seduced into formulating analytically deductive, prescriptive sets of doctrines that offer poor hope and worse guidance" (ibid., 87).

Beyerchen is less persuasive in his analysis of the "remarkable trinity" described at the end of the first chapter of Book I of *On War,* however. He makes a great deal of Clausewitz's supposed image of a theory floating amid the forces of emotion, courage and talent, and reason generated, respectively, by the people, the commander-in-chief, and the government "like an object suspended between three magnets." In other words, it is like a pendulum placed at the center of a triangular arrangement of three magnets, whose influence on the swinging pendulum produces irregular and therefore unpredictable motion. Clausewitz, Beyerchen speculates, may indeed have witnessed such an experiment before adopting the image as an emblem of the nonlinear nature of major armed conflict (ibid., 70–72).

The associated text from *On War,* however, makes Beyerchen's interpretation difficult to accept. For Clausewitz, the object suspended between the three poles of force is not war or the dynamics of war, but a theory of war. Its relationships to the forces of emotion, courage and talent, and reason, as embodied in the people, the commander-in-chief, and the government, Clausewitz is arguing, will change in proportion to variations in the character of these forces. The net effect of these variations, however, is not unpredictable movement, but a theory that remains in a position of "balance between these three tendencies" in spite of any differences in the values of the variables (*On War* [Howard/Paret], I/1: 89). A full discussion of the meaning and purpose of Clausewitz's famous analogy is reserved for the next chapter. For present purposes, enough has been said to indicate that in this particular instance, Beyerchen's argument is flawed.

In his introduction, Beyerchen warns that "an approach through nonlinearity does not make other reasons for difficulty in understanding *On War* evaporate" ("Nonlinearity," 61). This is to say that recognition of nonlinearity in Clausewitzian theory does not resolve all issues of major importance. Insofar as the moral aspect of war is concerned, nonlinearity is a useful concept because it facilitates understanding of the characteristics of war that make high command so difficult. But Beyerchen's findings are not helpful when it comes to the other side of the equation, namely, the characteristics of mind and spirit required of a commander-in-chief to problem-solve effectively in the face of the nonlinear dynamics of war, which Clausewitz also addresses. Fortunately, however, key elements of this matter—questions relating to the nature of decision-making under difficult conditions and how doing so effectively is learned—have been the subject of illuminating work in another scientific field.

Insofar as command decision is concerned, Clausewitz believes that the antidote to the uncertainty of war, which is largely the product of the inherent nonlinearity of major armed conflict, is the action of an unconscious intelligence.

His theoretical effort, therefore, is thus directed toward the improvement of unconscious as well as conscious cognition through the study of the past. "Action," Clausewitz insists, "can never be based on anything firmer than instinct, a sensing of the truth" (*On* War [Howard/Paret], I/3: 108).[71] The kind of bold decision-making that is essential to success in war, he declares, involves "rapid, only partly conscious weighing of the possibilities" (ibid., III/6: 192). He further argues that "the man of action must at times trust in the sensitive instinct of judgment, derived from his native intelligence and developed through reflection, which almost unconsciously hits on the right course" (ibid., III/14: 213). And "almost everything that happens in war," he maintains, is *"through the hidden processes of intuitive judgment"* (ibid., VI/8: 389; italics in original). Clausewitz is convinced that no great commander has acted in terms of clear logical paths—"when all is said and done," he observes, "it really is the commander's *coup d'oeil,* his ability to see things simply, to identify the whole business of war completely with himself, that is the essence of good generalship" (ibid., VIII/1: 578).[72] Clausewitz connects fear and inaction, which implies that fear disables intuition.[73] History of the kind he had described in Book II—that is, reenactment of command decision in the past—Clausewitz maintains, is the best preceptor of the "insights, broad impressions, and flashes of intuition" that are the main sources of sound decision (ibid., III/5: 185).

Clausewitz's views on the critical importance of unconscious cognition, the factors that interfere with it, and the means by which its performance can be improved have been substantiated to a remarkable degree by scientific research. In his 1997 book *Hare Brain, Tortoise Mind: How Intelligence Increases When You Think Less,* Guy Claxton, the professor of learning sciences at the University of Bristol Graduate School of Education, summarizes the findings of recent studies concerned with the nature and function of the human mind. This field of inquiry is known as "cognitive science," which he defines as "an alliance of neuroscience, philosophy, artificial intelligence and experimental psychology."[74] Within this broad subject area, Claxton focuses on matters pertaining to the restrictive effects of language on thinking, the shortcomings of deliberate reasoning with respect to problems that pose high levels of uncertainty, the nature of intuition and its role in decision-making, and the form of education best suited to the promotion of a strong capacity for effective decision-making under conditions of uncertainty. Claxton did not, it seems, consider the possibility that his work would be applicable to the psychology of high command and military education.[75] The congruence of his conclusions with those of Clausewitz is nonetheless very strong. And like Clausewitz, as well as Peirce, Collingwood, and Wittgenstein, Claxton's point of departure is the limitations of language as a medium of communicating certain things.

Deliberate thought, Claxton observes, is based on language. Language gives specific names to objects and actions, but the meaning of these names corresponds to attributes of objects and actions that are essential but not necessarily all significant. The great utility of language is that under many common circumstances, its inexactitude is outweighed by the power conferred by understandings of particular phenomena in terms that are general enough to facilitate useful division—that is, analysis—or combination—that is, synthesis. Complex social situations make it important for people to be able to "deconstruct responses to hybrid situations into familiar parts, and to be able to construct responses to hybrid situations by putting together different facets of different scripts. . . . This carving of scenarios into recombinable 'concepts' is basically, the ability conferred by language." Furthermore, Claxton says, "Language, and the ways of knowing which it affords, liberates; but it comes with snares of its own. Although it allows us to learn from the experience of others, and to segment and recombine our own knowledge in novel ways, it creates a different kind of rigidity" (*Hare Brain*, 43, 46).

Much that a person knows about the nature and dynamics of phenomena, Claxton maintains, cannot be described in words. Many facets of personal knowledge take the form of unconscious thought, such as a sense of "how things work" or "who can and who cannot be trusted" or simply "a hunch" that a particular situation requires this as opposed to that solution. Thus, "some of what we know is readily rendered into words and propositions; and some of it is not. Some of our mental operations are available to consciousness; and some of them are not. When we think, consciously and articulately, therefore, we are not capturing accurately all that is going on in the mind" (ibid., 91). Unconscious thought is no less important than conscious thought, especially when a person is confronted by situations that are "intricate, shadowy or ill defined" (ibid., 3). In such cases, effective decision-making depends upon the power of the unconscious mind to "detect, learn and use intricate patterns of information which deliberate conscious scrutiny cannot even *see*, under favourable conditions, let alone register and recall" (ibid., 25; italics in original).

Claxton calls the "intelligent unconscious" the "undermind" (ibid., 7). When the product of the undermind is registered consciously, it becomes intuition (ibid., 49). Intuition is not infallible, but it can augment and in certain instances surpass the power of conclusions that are the product of conscious thought. Recourse to intuition as the basis of decision-making is favored when either information or time is insufficient for deliberation. Use of intuition is disfavored, however, "when people feel threatened, pressurized, judged or stressed" (ibid., 75–76). This is because "when self-esteem is at stake, delicate unconscious forms of information and intelligence seem to be disabled or dismissed, and the way we act becomes clumsy and coarse" (ibid., 118). Or, in other words, "the more self-conscious we are—the more fragile our identity—the more we shut down the undermind" (ibid., 128). The good intuitive, therefore, is not just "the person who is ready, willing and able to make a lot out of a little" (ibid., 72). Quoting Malcolm Westcott, a professor at Vassar College, Claxton notes that people who are capable of acting on intuitive impulses are also those who can *explore uncertainties and entertain doubts far more than [others], and they live with these doubts and uncertainties without fear*" (ibid., 73–74; italics in original).

The capacity to deploy intuition effectively is wisdom, which Claxton defines as "good judgment in hard cases" (ibid., 191). And "hard cases," he observes,

> are those where important decisions have to be made on the basis of
> insufficient data; where what is relevant and what is irrelevant are not
> clearly demarcated; where meanings and interpretation of actions and
> motives are unclear and conjectural; where small details may contain
> vital clues; where the costs and benefits, the long-term consequences,
> may be difficult to discern; where many variables interact in intricate
> ways. (Ibid., 191)

It is for these reasons that "wisdom does not search the rule-book for templates and generalities that the situation can be forced to fit. It tends to go back to the moral and human basics" (ibid., 190). To be wise requires a person to entertain doubt while retaining the "freedom to act—sometimes quickly and decisively" (ibid., 195). Wisdom is, above all, "practical, dealing directly with 'matters relating to life and conduct.'" (ibid., 189).

The improvement of wisdom, Claxton insists, is not promoted by teaching that encourages students to use established techniques of problem-solving to find what are regarded as correct solutions. "If you are always fed a diet of problems that have been neatened up and graded you are deprived of the opportunity to develop those slow intuitive ways of knowing that are designed precisely to work best in situations that are untidy, foggy, ill conceived" (ibid., 222). In contrast, "interacting with complex situations without trying to figure them out can deliver a quality of understanding that defies reason and articulation" (ibid., 203). One cannot improve the power of unconscious thought "through earnest manipulation of abstraction"; instead, it requires "leisurely contemplation of the particular" (ibid., 173). In the latter case, a person must "extract patterns from experience, without necessarily being able to say what they are" (ibid., 220–221). Or, to put it in a way that connects this thought to others previously explained, the purpose of a properly conceived problem-solving exercise is not to find a solution, but to develop a person's unconscious mind by having him grapple with the emotional challenge of having to formulate action while in a state of extreme doubt and in spite of fear of failure.

Claxton's findings indicate that Clausewitz's concept of reenacting the psychological circumstances of supreme command under difficult conditions, as defined by the historical record and theoretical surmise, embody insights into the nature of human intelligence that have been confirmed by cognitive science. Moreover, the approach to learning called for by Clausewitz's synthesis of history and theory comports with conclusions reached by modern educators. There are two important implications for military education. First, to improve intuition—the key cognitive agent with respect to the negotiation of war as an unstable nonlinear phenomenon—one must address the education of the intelligent unconscious and the moral nature of strategic decision-making. And second, the theoretical description of system dynamics is not an appropriate basis for officer education because such a method is inherently incapable of engaging the undermind, whereas reenactment, when properly constructed, offers a promising method of doing so. In the undated note that was probably written in 1827, Clausewitz observes that "it is a very difficult task to construct a scientific theory for the art of war" because ". . . it deals with matters that no permanent law can provide for" (*On War* [Howard/Paret], 71). On the strength of the foregoing discussion, it would seem that he nevertheless succeeded.

4

Imagining High Command and Defining Strategic Choice

COMPREHENSION OF THREE PAIRS OF CONCEPTS is essential to understanding *On War*. The first pair has to do with those factors that Clausewitz believed define the proper use of active armed force in international relations, namely, absolute war and genius. The second pair is concerned with Clausewitz's thinking about history and theory and the relationship between them. And the third pair is about Clausewitz's views on defense and attack. In this chapter we will look at each pair of themes in turn. Such an analysis cannot of course address every issue raised in *On War*. As Clausewitz himself observes, "No analytical system can ever be explored exhaustively" (*On War*, VII/1: 523).[1] But it should enable readers to follow the author's ways of thinking about matters that he believed were of the utmost theoretical and national importance and provide a sound point of departure for the productive reading of Clausewitz's great study of armed conflict.

Absolute War and Genius

In the first chapter of Book I, Clausewitz defines war as "an act of force to compel our enemy to do our will." He defines extreme or absolute war as a phenomenon in which the behavior of one or both combatants is driven by the need to maximize the use of force without restriction. There are, according to Clausewitz, three factors that promote the maximization of violence. First, war consists of using force, and so long as compulsion of a resisting opponent is considered necessary, there is no logical limit on its use. Second, each side is motivated by fear that a failure to use the maximum possible degree of force might result in defeat. And third, both sides are likely to attempt to win by surpassing the other with respect to the magnitude of force used, resulting in competitive actions that progressively increase violence to the maximum level (ibid., I/1: 75–77).

Chapter Four

Clausewitz insists, however, that all war cannot be understood in terms of its absolute form. In the real world, he observes, three basic moderating factors act against the maximization of violence. First, particular circumstances external to war—namely, the character of national leadership, national institutions, and national and international political situations on both sides—create conditions that restrain the tendency to use or maximize the use of violence. Second, because parties to a war are aware that the outcome cannot be determined by a single blow, resources on both sides are reserved for later employment, the effect of which is to diminish the use of force on any one occasion. And third, even a state defeated in a major engagement can regard the loss as no more than a temporary setback whose effects will be reversed at some indeterminate time in the future, which will promote a willingness in the short run to come to terms as an alternative to maximizing military effort (ibid., I/1: 78–80).

Clausewitz then observes that whenever the propensity to maximize violence is counterbalanced by the moderating factors just described, the incentives for action will be provided by political objectives. The relationship between political objectives and military effort is in general proportional. That is, small political objectives will elicit less military effort, while large political objectives will call for more military effort. If the latter case involves the survival of one or both sides, the largeness of the military effort on either or both sides may produce what is in effect the maximization of violence. Thus, Clausewitz concludes, "wars . . . can have all degrees of importance and intensity, ranging from a war of extermination down to simple armed observation" (ibid., I/1: 81).

To sum up thus far, absolute war as an abstraction is war in which the maximization of violence is determined solely by the dynamics of the use of force. But the maximization of violence can also be brought about as a matter of policy. As Clausewitz declares in his opening remarks, "the maximum use of force is in no way incompatible with the simultaneous use of the intellect" (ibid., I/1: 75). And later he writes,

> The more powerful and inspiring the motives for war, the more they
> affect the belligerent nations and the fiercer the tensions that precede
> the outbreak, the closer will war approach its abstract concept, the
> more important will be the destruction of the enemy, the more
> closely will the military aims and the political objects of war coincide,
> and the more military and less political will war appear to be. (Ibid.,
> I/1: 87–88)

Clausewitz insists, moreover, that "among the contingencies for which the state must be prepared is a war in which every element calls for policy to be eclipsed by violence" (ibid., I/1: 88). Thus, while Clausewitz characterizes absolute war as an abstraction, and war that is less than absolute as real, *there is no question but that he recognizes that war that involves the unrestrained use of violence can occur and thus presumably is also real.*

These concepts—absolute war as an abstraction and real war that amounts to absolute war—are less problematical than may be apparent at first glance. The onset of hostilities in both cases is determined by politics/policy, namely, the attacker's objective of imposing his will and the defender's objective of resisting such action. The difference between the two is with respect to what happens afterward. In absolute war as an abstraction, action throughout is determined by the maximization of the use of force without respect to politics/policy. In (real) absolute war, the maximization of violence might occur and even govern behavior for an extended period, but it does not determine the course of the entire conflict. While the maximization of violence can overshadow all other considerations for a time, political/policy factors will in general exert a controlling influence. That said, it is also the case that the character of the fighting can affect political/policy factors reciprocally:

War is a pulsation of violence, variable in strength and therefore variable in the speed with which it explodes and discharges its energy. War . . . always lasts long enough for influence to be exerted on the goal and for its own course to be changed in one way or another— long enough, in other words, to remain subject to the action of superior intelligence. If we keep in mind that war springs from some political purpose, it is natural that the prime cause of its existence will remain the supreme consideration in conducting it. That, however, does not imply that the political aim is a tyrant. It must adapt itself to its chosen means, a process which can radically change it; yet the political aim remains the first consideration. Policy, then, will permeate all military operations, and, in so far as their violent nature will admit, it will have a continuous influence on them. (Ibid., I/1: 87)[2]

Clausewitz explains the defining characteristics of absolute war as an abstraction from the point of view of the attacker. These are, first, that the enemy's "fighting forces must be destroyed"; second, that the enemy's "country must be occupied"; and third, that the enemy's "government and its allies" must be "driven to ask for peace" and the population "made to submit" (ibid., I/2: 90). He then argues that (real) absolute war must take into account the strength of defender resistance, which can defeat the invasion or discourage its consummation through protracted resistance. "Since war is not an act of senseless passion," Clausewitz observes,

> but is controlled by its political object, the value of the object must determine the sacrifices to be made for it in *magnitude* and also in *duration*. Once the expenditure of effort exceeds the value of the political object, the object must be renounced and peace must follow. (Ibid., I/2: 92; italics in original)

Determined resistance by the defender, in other words, can cause the invader to alter its political/policy objectives. In war, Clausewitz says, "the original political objects can greatly alter during the course of the war and may finally change entirely *since they are influenced by events and their probable consequences*" (ibid.; italics in original).

Clausewitz maintains that if the defender makes "use of every means available for pure resistance," and if this is sufficient "to *balance* any superiority the opponent may possess," then in the end the attacker's "political object will not seem worth the effort it costs" and the attacker will decide to "renounce his policy." "It is evident that this method," he observes, "wearing down the enemy, applies to the great number of cases where the weak endeavor to resist the strong." Whereas the objective of an attacker when waging real war—which could be absolute or less than absolute—is to overcome the defender's powers of resistance through decisive battle, the objective of the defender is to preserve these powers of resistance and to weaken or exhaust the attacker through the protraction of the war. Clausewitz puts it succinctly: "The policy with a positive purpose calls the act of destruction into being; the policy with a negative purpose waits for it." Waiting, however, need not be entirely passive, even if the defending army has been shattered. With respect to such extreme conditions, Clausewitz's mention of the "use of every means available" undoubtedly refers to People's War (ibid., I/2: 94, 98; italics in original).

Clausewitz reserves his main explanation of People's War for Book VI, but he alludes to its importance as early as the first chapter of Book I. There he notes that the resources available to a state defending its own territory are "the fighting forces proper, the country, with its physical features and population, and its allies," and that the physical features and population are more than just the source of the armed forces proper, being themselves "an integral element among the factors at work in war—though only that part which is the actual theater of operations, or has a notable influence on it" (ibid., I/1: 79). In the second chapter of Book I, he repeatedly states that further explanation of important points related to attack and defense—which almost certainly refers to People's War—would come later (ibid., I/2: 92–94, 98–99).

In Book VI, Clausewitz defines People's War as state-sponsored insurrection by armed civilians against an invader in support of action by the regular army and the regular forces of allies. A general insurrection, he explains, is "simply another means of war" that had been brought about

> as an outgrowth of the way in which the conventional barriers have
> been swept away in our life-time by the elemental violence of war. It
> is, in fact, a broadening and intensification of the fermentation
> process known as war. The system of requisitioning, and the
> enormous growth of armies resulting from it and from universal
> conscription, the employment of militia—all of these run in the same
> direction when viewed from the standpoint of the older, narrower
> military system, and that also leads to the calling out of the home
> guard and arming the people. (Ibid., VI/26: 479)

People's War, according to Clausewitz, is thus on the leading edge of the expansion of the domain of violence in war and can escalate the ferocity with which violence is used. It was precipitated by the advent of offensive (real) absolute war, and as such is a defensive variant of (real) absolute war.

The ultimate aim of defensive (real) absolute war is the preservation of national sovereignty. Under certain conditions, this can be achieved through the destruction of the invading army, but this is not essential. When the balance of forces heavily favors the attacker, ruling out such a course, the defender can resort to People's War as a way of fighting a (real) absolute war. The promotion of attacker discouragement by the consequent increases in the costs of invasion and occupation will result in the defender's survival in spite of attacker victories in all major battles. War in which the destruction of the enemy military forces is not the objective, Clausewitz maintains, is by definition a limited war (ibid., VIII/8: 613). But as has just been explained, such a limited war in Clausewitzian terms can be fought as a People's War. In Clausewitz's theoretical universe, therefore, a defensive (real) absolute war can also be a limited war.

An understanding of Clausewitz's conception of People's War and its implications is critical to making sense of his famous closing section to the first chapter of
Book I. Here he argues that the dominant tendencies of war make up "a remarkable trinity" consisting of the emotional character of the general population, the
ability of the commander-in-chief to direct the armed forces effectively, and the
political will and acumen of the government (ibid., I/1: 89). In the case of executing the most extreme form of defensive (real) absolute war—that is, People's
War—in response to offensive (real) absolute war, success will require three
things: first, passionate popular support of the political/policy objectives of the
government; second, a commander-in-chief capable of making sound decisions
in the face of highly complex and very difficult strategic circumstances brought
about by the occupation of home territory and the inability of the regular army
to defeat the invading army; and third, a government committed to victory at
any cost, even in the face of fomenting civil disorder, whose potential long-run
political dangers are bound to be considerable. In the case of the opposite extreme—that is, armed observation—popular support of military action is irrelevant to the outcome; decision-making on the part of the commander-in-chief
will not be complicated by enemy occupation of home territory or the necessity
of fighting; and the government will not be confronted with a choice between
national survival, on the one hand, and large and potentially highly disruptive
changes in internal political arrangements, on the other. A war that falls some-
where in between these two boundaries will set proportionally different values
for each of the three variables.

Clausewitz thus argues that a universally applicable theory of war must take
into account that each of the three variables and their interrelationships can
vary widely. Such a general theory of war, therefore, has to be constructed in a
way that ensures that changes in the values of the three elements of the trinity
will not invalidate the basic propositions of that theory. The "three tendencies,"
Clausewitz explains,

> are like three different codes of law, deep-rooted in their subject and
> yet variable in their relationship to one another. A theory that ignores
> any one of them or seeks to fix an arbitrary relationship between them
> would conflict with reality to such an extent that for this reason alone
> it would be totally useless.

"Our task therefore," he concludes, "is to develop a theory that maintains a
balance between these three tendencies, like an object suspended between three

magnets" (ibid.). The point of this metaphor is further explained in the next chapter. Here Clausewitz enjoins his readers to

> bear in mind how wide a range of political interests can lead to war, or think for a moment of the gulf that separates a war of annihilation, a struggle for political existence, from a war reluctantly declared in consequence of political pressure or of an alliance that no longer seems to reflect the state's true interests. Between these two extremes lie numerous gradations. If we reject a single one of them on theoretical grounds, we may as well reject them all, and lose contact with the real world. (Ibid., I/2: 94)[3]

Clausewitz notes that his description of a theory that is applicable to all cases without exception would be given in Book II. This portion of *On War* is largely devoted to his conception of historical reenactment, and will be dealt with in due course.

The purpose of Clausewitz's "remarkable trinity" is to illustrate the requirements that must be met in order to formulate a valid general theory of war. *The elements of the trinity do not, however, define Clausewitz's main subject.* Two of the three elements—"the people" and "the government"—are entities concerned with matters connected to but nonetheless external to war—that is, politics/policy. The "passion" of the people can result in political action in the form of armed support for regular military forces, while the "reason" of the government stipulates the political ends to which force of any kind will be put. It is "the commander and his army" that constitute, respectively, the executor and the main instrument of force, and thus they are the fundamental factors of war. The direction of the main instrument of force, and any support by armed civilians, is the sole responsibility of the commander. Clausewitz thus focuses his analytical effort on the commander-in-chief. Effective direction of the war, he says, depends upon his "courage and talent." These attributes are unquantifiable and infinitely variable. Clausewitz thus establishes their critical importance to success in war through qualitative description.

In the first chapter of Book I, Clausewitz describes the three main factors that can cause fighting during war to be suspended, thus making decisions during war by the commander-in-chief "a matter of assessing probabilities" rather than simply an act of will to maximize the use of force. The first is the "very great" superiority of the defense over the attack (to be described in more detail later), which can discourage further action on the part of the attacker. The second is "imperfect knowledge of the situation," which can cause the commander-in-chief "to suppose that the initiative lies with the enemy when in fact it remains with him." And the third is the very high degree to which outcomes in war are affected by chance, which means that the initiation of action by the commander-in-chief requires courage in the face of the possibility that failure can occur even under the most favorable conditions (ibid., I/1: 84–86).

For Clausewitz, the issue of chance in war is crucial. It is the play of chance that makes it impossible to base decision-making by the commander-in-chief on reason alone. Thus Clausewitz argues that

> absolute, so-called mathematical, factors never find a firm basis in military calculations. From the very start there is an interplay of possibilities, probabilities, good luck and bad that weaves its way throughout the length and breadth of the tapestry. In the whole range of human activities, war most closely resembles a game of cards. (Ibid., I/1: 86)

The antidote to fear promoted by the possibility of chance-induced failure is a form of psychological strength that exists outside the boundaries of deliberate thought. This antidote is essential if action is to even be initiated. "Although our intellect always longs for clarity and certainty," Clausewitz explains,

> our nature often finds uncertainty fascinating. It prefers to day-dream in the realms of chance and luck rather than accompany the intellect on its narrow and tortuous path of philosophical inquiry and logical deduction only to arrive—hardly knowing how—in unfamiliar surroundings where all the usual landmarks seem to have disappeared. Unconfined by narrow necessity, it can revel in a wealth of possibilities; which inspire courage to take wing and dive into the element of daring and danger like a fearless swimmer into the current. (Ibid.)

Clausewitz is convinced that theory must address the fact that commanders-in-chief make difficult decisions in war largely on the basis of certain emotions. Theory, he declares,

> must also take the human factor into account, and find room for courage, boldness, even foolhardiness. The art of war deals with living and with moral forces. Consequently, it cannot attain the absolute, or

certainty; it must always leave a margin for uncertainty, in the greatest things as much as in the smallest. With uncertainty in one scale, courage and self-confidence must be thrown into the other to correct the balance. The greater they are, the greater the margin that can be left for accidents. Thus courage and self-confidence are essential in war, and theory should propose only rules that give ample scope to these finest and least dispensable of military virtues in all their degrees and variations. Even in daring there can be method and caution; but here they are measured by a different standard. (Ibid.)

"Such is war," Clausewitz concludes, "such is the commander who directs it, and such the theory that governs it" (ibid.).

The observations on the nature of strategic decision-making in the first two chapters of Book I are introductory. Clausewitz undertakes a comprehensive and systematic exposition of the entire question of what it takes to be an effective commander-in-chief in Chapters 3 through 8. In the third chapter, entitled "On Military Genius," Clausewitz surveys "all those gifts of mind and temperament that in combination bear on military activity," which "taken together, *constitute the essence of military genius.*"[4] For Clausewitz, genius is not a single attribute, but a complex entity whose constituent elements can vary widely from one case to the next. "Genius," he observes, "consists in a harmonious combination of elements, in which one or the other ability may predominate, but none may be in conflict with the rest." In Clausewitz's mind, military genius is not supreme capability—that is, talent that lacks "measurable limits"—but highly developed competence in the performance of certain difficult tasks. The defining quality of genius is the kind of sophisticated intelligence produced by advanced societies. "With every people renowned in war," Clausewitz declares, "the greatest names do not appear before a high level of civilization has been reached" (ibid., I/3: 100–101; italics in original).

Clausewitz recognizes that the two basic attributes of the soldier common to both primitive and advanced societies are nonintellectual, namely, physical courage and endurance in the face of privation and suffering. But war at the level of supreme command waged by an advanced society is a "region dominated by the *powers of intellect.*" This is because strategic leadership functions within "the realm of uncertainty" (ibid., I/3: 101; italics in original):

Since all information and assumptions are open to doubt, and with chance at work everywhere, the commander continually finds that things are not as he expected. . . . During an operation decisions have usually to be made at once: there may be no time to review the situation or even to think it through. Usually, of course, new information and reevaluation are not enough to make us give up our intentions: they only call them into question. We now know more, but this makes us more, not less uncertain. (Ibid., I/3: 102)

"If the mind is to emerge unscathed from this relentless struggle with the unforeseen," Clausewitz concludes, "two qualities are indispensable: *first, an intellect that, even in the darkest hour, retains some glimmerings of the inner light which leads to truth; and second, the courage to follow this faint light wherever it may lead*" (ibid.; italics in original).

Clausewitz explains that his notion of an intellect informed by an "inner light" is a form of the French term *coup d'oeil*, which means a comprehensive view made in an instant. Its usage is normally associated with visual observation and instantaneous assessment of enemy dispositions as the basis of deploying cavalry—that is to say, with tactics. Clausewitz argues that *coup d'oeil* at the strategic level can be defined as a mental rather than a visual act of observation and assessment of general circumstances, which he calls the action of "the inward eye." "Stripped of metaphor [that is, inward eye] and of the restrictions imposed on it by the phrase [that is, *coup d'oeil*]," Clausewitz observes, "the concept merely refers to the quick recognition of a truth that the mind would ordinarily miss or would perceive only after long study and reflection." For Clausewitz, *coup d'oeil* is the military term for what is otherwise known as intuition (ibid.).

Clausewitz is convinced that this kind of intuition cannot be brought to bear unless determination overcomes fear. "The role of determination," Clausewitz explains, "is to limit the agonies of doubt and the perils of hesitation when the motives for action are inadequate." Determination, in turn, is the product of courage. Courage at the level of strategic decision-making, Clausewitz argues, is "the courage to accept responsibility, courage in the face of a moral danger." High intelligence does not generate such courage automatically. "Intelligence alone," Clausewitz notes, "is not courage; we often see that the most intelligent people are irresolute" (ibid., I/3: 102–103). To produce the courage needed, high intelligence has to be deployed in a certain way:

Determination, which dispels doubt, is a quality that can be aroused only by the intellect, and by a specific cast of mind at that. More is required to create determination than a mere conjunction of superior insight with the appropriate emotions. Some may bring the keenest brains to the most formidable problems, and may possess the courage to accept serious responsibilities; but when faced with a difficult situation they still find themselves unable to reach a decision. Their courage and their intellect work in separate compartments, not together; determination, therefore, does not result. It is engendered only by a *mental act;* the mind tells man that boldness is required, and thus gives direction to his will. This particular cast of mind, which employs the fear of *wavering* and *hesitating* to suppress all other fears, is the force that makes strong men determined. (Ibid., I/3: 103; italics in original)

Clausewitz completes his delineation of the intellectual components of genius with a brief description of what he calls "*presence of mind.*" This attribute, he writes, "must play a great role in war, the domain of the unexpected, since it is nothing but an increased capacity of dealing with the unexpected." Clausewitz adds that "whether this splendid quality is due to a special cast of mind or to steady nerves depends on the nature of the incident, but neither can ever be entirely lacking" (ibid., I/3: 103–104; italics in original).

The three intellectual components of genius—*coup d'oeil,* determination, and presence of mind—address the problems posed by uncertainty. A second major challenge to the psychology of the commander-in-chief is the resistance of his subordinates to direction, the strength of which is proportional to the fortunes of the army on campaign:

So long as a unit fights cheerfully, with spirit and élan, great strength of will is rarely needed; but once conditions become difficult, as they must when much is at stake, things no longer run like a well-oiled machine. The machine itself begins to resist, and the commander needs tremendous will-power to overcome this resistance. The machine's *resistance* need not consist of disobedience and argument, though this occurs often enough in individual soldiers. It is the impact of the ebbing of moral and physical strength, of the heart-rending spectacle of the dead and wounded, that the commander has to withstand—first in himself, and then in all those who, directly or indirectly, have entrusted him with their thoughts and feelings, hopes and fears. As each man's strength gives out, as it no longer responds to his will, the inertia of the whole gradually comes to rest on the commander's will alone. . . . The burdens increase with the number of

men in his command, and therefore the higher his position, the
greater strength of character he needs to bear the mounting load.
(Ibid., I/3: 104–105; italics in original)[5]

Thus genius, for Clausewitz, is not only defined by intellectual power, but
by personal temperament, which conditions the commander-in-chief's relationships with subordinates under difficult circumstances. The temperament
aspect of military genius as addressed by Clausewitz has four elements: energy,
staunchness, strength of character, and firmness. Energy is the product of emotion, and the most important aspect of it for Clausewitz is ambition. "We may
well ask," Clausewitz writes, "whether history has ever known a great general
who was not ambitious; whether, indeed, such a figure is conceivable." Staunchness means the ability to withstand high degrees of psychological shock (ibid., I/
3: 105). Strength of character, Clausewitz explains, is

> an emotion which serves to balance the passionate feelings in strong
> characters without destroying them, and it is this balance alone that
> assures the dominance of the intellect. The counterweight we mean is
> simply the sense of human dignity, the noblest pride and deepest need of
> all: the urge *to act rationally at all times.* (Ibid., I/3: 106; italics in original)

Firmness means having the will to pursue a course of action in spite of setbacks based upon instruction from one's broad understanding of the nature of
things. "Only those general principles and attitudes," Clausewitz maintains,
"that result from clear and deep understanding can provide a *comprehensive*
guide to action. It is to these that opinions on specific problems should be an-
chored" (ibid., I/3: 108; italics in original).

Clausewitz concludes his descriptive analysis of genius with a discussion of
a mental attribute of genius that is not related either to intellectual power as
such or to temperament. This is a capacity to comprehend the military significance of topography:

> A commander-in-chief . . . must aim at acquiring an overall
> knowledge of the configuration of a province, of an entire country.
> His mind must hold a vivid picture of the road-network, the river-
> lines and the mountain ranges, without ever losing a sense of his
> immediate surroundings. . . . With a quick, unerring sense of locality
> his dispositions will be more rapid and assured; he will run less risk of
> a certain awkwardness in concepts, and be less dependent on others.

We attribute this ability [Clausewitz concludes] to the
imagination; but that is about the only service that war can demand
from this frivolous goddess, who in most military affairs is liable to do
more harm than good. (Ibid., I/3: 110)

In his summation of his chapter on genius, Clausewitz again insists upon
the critical importance of intellectual capability. Although he had divided his
definition of genius into matters of mind and matters of temperament, his discussion of the latter is such as to relate temperament, and even a separate ability
to suppress fear, to intellectual action. He thus observes that in his "review of the
intellectual and moral powers that human nature needs to draw upon in war.
The vital contribution of intelligence is clear throughout. No wonder then, that
war, though it may appear to be uncomplicated, cannot be waged with distinction except by men of outstanding intellect." While Clausewitz concedes that
high levels of intellectual powers are required at all levels of command, he maintains that "history and posterity reserve the name of 'genius' for those who have
excelled in the highest positions—as commanders-in-chief—since here the demands for intellectual and moral powers are vastly greater" (ibid., I/3: 110–111).

Clausewitz explains the nature of those demands as follows. "To bring a
war, or one of its campaigns," he writes, "to a successful close requires a
thorough grasp of national policy. On that level strategy and policy coalesce: the
commander-in-chief is simultaneously a statesman." That said, Clausewitz
warns that a general still has to be capable of deploying an army effectively,
which means that although "a commander-in-chief must also be a statesman,
. . . he must not cease to be a general." The execution of both roles requires solution of contingent problems of great complexity and difficulty. "Circumstances
vary so enormously in war, and are so indefinable," he argues, "that a vast array
of factors has to be appreciated—mostly in the light of probabilities alone. The
man responsible for evaluating the whole must bring to his task the quality of
intuition that perceives the truth at every point." Such intuition, Clausewitz explains, "is a sense of unity and a power of judgment raised to a marvelous pitch
of vision, which easily grasps and dismisses a thousand remote possibilities
which an ordinary mind would labor to identify and wear itself out in so doing"
(ibid., I/3, 111–112). Intuition, moreover, has to be buttressed by moral power to
produce action:

Truth in itself is rarely sufficient to make men act. Hence the step is
always long from cognition to volition, from knowledge to ability. The

most powerful springs of action in men lie in his emotions. He derives his most vigorous support, if we may use the term, from that blend of brains and temperament which we have learned to recognize in the qualities of determination, firmness, staunchness, and strength of character. (Ibid., I/3, 112)

In addition to attributes of mind and temperament, the capability of a commander-in-chief is heavily influenced by comprehension of certain unquantifiable characteristics inherent to war and the ability to come to terms with their effects. Clausewitz describes these factors in the four chapters that follow the chapter on genius. The first is awareness of physical danger on the battlefield and its negative effects on the powers of quick decision (ibid., I/4: 113–114). The second is the debilitating physical demands of a campaign and their effects on the body and spirit of the commander-in-chief (ibid., I/5: 115–116). The third is the problem of incomplete and misleading information and the challenges this presents for a commander's mental equilibrium (ibid., I/6: 117–118). And the fourth is what Clausewitz called "friction," which refers to all the forces that interfere with an army's responsiveness to direction by the commander-in-chief (ibid., I/7: 119–121). In the last chapter of Book I, Clausewitz groups danger, physical exertion, intelligence, and friction into a single category that he called "general friction" (ibid., I/8: 122).

Clausewitz offers incidental observations on friction—which undoubtedly apply to the concept of general friction as well—that relate the concept to the faculties of command that constitute genius. The ability to surmount friction, he maintains, requires "the development of instinct and tact," which constitute "a form of judgment" capable of dealing with "an area littered with endless minor obstacles." "As with a man of the world," Clausewitz says,

instinct becomes almost habit so that he always acts, speaks, and moves appropriately, so only the experienced officer will make the right decision in major and minor matters—at every pulsebeat of war. Practice and experience dictate the answer: "this is possible, that is not." So he rarely makes a serious mistake, such as can, in war, shatter confidence and becomes extremely dangerous if it occurs often. (Ibid., I/7: 120–121)

In the last chapter of Book I, Clausewitz observes that the effects of friction can be reduced by combat experience. When this is not "readily available," maneuvers and the advice of foreign officers with experience can offer a measure of compensation (ibid., I/8: 122–123).

By the end of Book I, Clausewitz has defined three main components of decision-making competence on the part of the commander-in-chief: mind, temperament, and experience. He is convinced that the development of the appropriate mind and temperament—the constituent elements of military genius—is favored by intellectual curiosity, broad perspective, and emotional discipline. "If we then ask what sort of mind is likeliest to display the qualities of military genius," Clausewitz writes in his final paragraph of Chapter 3, "experience and observation will both tell us that it is the inquiring rather than the creative mind, the comprehensive rather than the specialized approach, the calm rather than the excitable head to which in war we would choose to entrust the fate of our brothers and children, and the safety and honor of our country" (ibid., I/3: 112). But these qualities are not enough—they need to be augmented and extended by experience of actual war. Clausewitz believes that it is possible to formulate a theory of war that will promote the operation of genius through the replication of the effects of experience. Theory of this kind addresses the improvement of intuitive as well as deliberate thought—that is, the education of the unconscious as well as of the conscious mind. He describes such a theory in the second book of *On War*.

History and Theory

Conventional military theory in Clausewitz's time was largely a matter of drawing lessons from the study of past campaigns and battles. Or, in other words, examination of the particular in the past was the source of general propositions that were supposed to be capable of providing useful instruction about the particular in the future. Clausewitz regarded such a procedure as problematical because he believed that every historical event contained unique episodes that had a critical effect on action and outcomes. Trying to apply such propositions could be dangerous for two different reasons. First, the generalization could so faithfully represent the dynamics of the past event as to make it inapplicable to any future event unless that event very closely replicated the conditions of the historical source, in which case the theory would lack the universality that Clausewitz considers essential. Or second, the generalization could universalize its applicability by ignoring or factoring out significant but unique aspects of the past event, at the cost of forfeiting any legitimate claim to historical validation.

In Book II, which is entitled "On the Theory of War," Clausewitz thus invents a genre of theory that differs fundamentally in form and function from convention. Instead of studying the particular in the past in order to elucidate the general, he establishes the validity of certain general propositions, which are then used to generate additions to the historical record in order to provide what could be regarded as a more complete representation of the dynamics of command decision. This combination of theory and history amounts to historical reenactment of strategic and tactical decision-making experience, which is to be followed by reflection on that experience. The two actions constitute a form of learning that approximates that which takes place in actual war of any kind, and thus satisfies Clausewitz's requirement for theory that is applicable to either (real) absolute or (less than absolute) real war. In addition to this innovative approach, Clausewitz maintains that history can be used to prove that a certain general proposition is valid for both kinds of armed conflict—specifically, proper analysis of the historical record can demonstrate beyond question that the defense is superior to the attack in (real) absolute as well as (less than absolute) real war. Clausewitz regards his theoretical approaches as scientific because their primary function is to make possible productive observation of particular experience—that is, to facilitate what can be called historical psychological experiment—rather than to serve as guides to decision-making.

In Book I, Chapter 4, Clausewitz in effect reenacts the psychological experience of a novice participating in his first engagement. "As we approach [the fighting]," he writes, assuming the perspective of such a person,

> the rumble of guns grows louder and alternates with the whir of
> cannonballs, which begin to attract his attention. Shots begin to strike
> close around us. We hurry up the slope where the commanding
> general is stationed with his large staff. Here cannonballs and bursting
> shells are frequent, and life begins to seem more serious than the
> young man had imagined. Suddenly someone you know is wounded;
> then a shell falls among the staff. You notice that some of the officers
> act a little oddly; you yourself are not as steady and collected as you
> were: even the bravest can become slightly distracted. Now we enter
> the battle raging before us, still almost like a spectacle, and join the
> nearest divisional commander. Shot is falling like hail, and the
> thunder of our own guns adds to the din. Forward to the brigadier, a

soldier of acknowledged bravery, but he is careful to take cover behind
a rise, a house or clump of trees. A noise is heard that is a certain
indication of increasing danger—the rattling of grapeshot on roofs
and on the ground. Cannonballs tear past, whizzing in all directions,
and musketballs begin to whistle around us. A little further we reach
the firing line, where the infantry endures the hammering for hours
with incredible steadfastness. The air is filled with hissing bullets that
sound like a sharp crack if they pass close to one's head. For a final
shock, the sight of men being killed and mutilated moves our
pounding hearts to awe and pity. (Ibid., I/4: 113)

Readers of this passage, Clausewitz observed, should be able to sense that in
the face of danger, "ideas are governed by other factors, that the light of reason
is refracted in a manner quite different from that which is normal in academic
speculation" (ibid.). And while the perspective of his exercise was tactical—that
is, concerned with the battlefield—there can be little doubt that he believed that
forces of a similar kind but even greater in magnitude affected the exercise of
command at the strategic level—that is, in the direction of campaigns to achieve
the political objectives of the war. On the battlefield, "ordinary qualities" of
equanimity, Clausewitz writes,

are not enough; and the greater the area of responsibility, the truer this
assertion becomes. Headlong, dogged or innate courage,
overmastering ambition, or long familiarity with danger—all must be
present to a considerable degree if action in this debilitating element is
not to fall short of achievements that in the study would appear as
nothing out of the ordinary. (Ibid., I/4: 114)

Clausewitz reserves description of what might constitute an instructive reenactment of the psychological conditions of strategic command decision,
which had to address far more difficult and complex issues, for Book II.

At the beginning of Book II, he defines the subject that theory is to address
through a process of eliminating matters that he regards as inessential. Clausewitz draws a sharp distinction between activities "*that are merely preparations for
war, and war proper.*" The former category consists of "the creation of the fighting forces, their raising, armament, equipment, and training," as well as "supply,
medical services, and maintenance of arms and equipment." He points out that
these issues need not be addressed by a general theory of war. War proper is essentially fighting, which has physical and moral components, the latter of which

exert a decisive influence. He then divides war proper into two levels: tactics, which is *"the use of armed forces in the engagement,"* and strategy, which is *"the use of engagements for the object of the war."* Tactics, in other words, is about decision-making with respect to how a battle is fought, whereas strategy is about decision-making with respect to how battles can be used to bring a war to the most favorable possible conclusion (ibid., II/1: 127–131; italics in original). Clausewitz ends his introductory definitional survey with an observation about existing theory:

> Anyone for whom all this is meaningless either will admit no
> theoretical analysis at all, or his intelligence has never been insulted by
> the confused and confusing welter of ideas that one so often hears and
> reads on the subject of the conduct of war. These have no fixed point
> of view; they lead to no satisfactory conclusion; they appear
> sometimes banal, sometimes absurd, sometimes simply adrift in a sea
> of vague generalization; and all because this subject has seldom been
> examined in a spirit of scientific investigation. (Ibid., II/1: 132)

Clausewitz begins the second chapter of Book II with a review of the history of military theory. He notes that the term "art of war" originally was concerned only with the preparation of military forces, and did not, therefore, consider "the use of force under conditions of danger, subject to constant interaction with an adversary, nor the efforts of spirit and courage to achieve a desired end." In the next stage, writers were concerned with siege warfare and theories of tactics, applying intellectual effort in an attempt to solve operational problems, but in ways that did not amount to much more than description of technique. For many years examination of the "actual conduct of war"—that is, "the free use of the given means, appropriate to each individual occasion"—was not considered a suitable subject for theory "but one that had to be left to natural preference." But the proliferation of military memoirs and histories generated controversies that called for resolution through reference to "basic principles and clear laws." The result was that theorists attempted to "equip the conduct of war with principles, rules, or even systems" (ibid., II/2: 133–134).

These efforts were compromised by the inability of theory so formed to deal with the "endless complexities" of war, which create an "irreconcilable conflict" between theory and practice. Discouraged by this outcome, military theorists, Clausewitz maintains, directed their "principles and systems only to physical matters and unilateral activity," so that they could "reach a set of sure and positive conclusions" (ibid., II/2: 134). They thus wrote about such matters as numerical superiority, supply, bases, and interior lines. All such writing, Clausewitz argues, was "objectionable":

> It is only analytically that these attempts at theory can be called advances in the realm of truth; synthetically, in the rules and regulations they offer, they are absolutely useless.
>
> They aim at fixed values; but in war everything is uncertain, and calculations have to be made with variable quantities.
>
> They direct the inquiry exclusively toward physical quantities, whereas all military action is intertwined with psychological forces and effects.
>
> They consider only unilateral action, whereas war consists of a continuous interaction of opposites. (Ibid., II/2: 136)

Theory of the kind just described relegates all unquantifiable aspects of the conduct of war to the action of genius, which is left uninvestigated because it is regarded as "beyond scientific control." Clausewitz is convinced, however, that the domain of genius covers the greater and much more important part of what constitutes the conduct of war, and that to ignore it while addressing lesser matters amounts to collecting the chaff and discarding the wheat. He regards such conduct with contempt. "Pity the soldier," Clausewitz writes, "who is supposed to crawl among these scraps of rules, not good enough for genius, which genius can ignore, or laugh at. No; what genius does is the best rule, and theory can do no better than show how and why this should be the case" (ibid.).

The scientific investigation of genius, however, poses enormous problems because of the difficulty of dealing with moral factors. As Clausewitz explains,

> Theory becomes infinitely more difficult as soon as it touches the realm of moral values. Architects and painters know precisely what they are about as long as they deal with material phenomena. Mechanical and optical structures are not subject to dispute. But when they come to the aesthetics of their work, when they aim at a particular effect on the mind or on the senses, the rules dissolve into nothing but vague ideas. (Ibid.)

Moral values, however, cannot be ignored in any discussion of the conduct of war because they are central. "Military activity," Clausewitz argues, "is never directed against material force alone; it is always aimed simultaneously at the moral forces which give it life, and the two cannot be separated" (ibid., II/2: 137).

Chapter Four

Clausewitz's solution to the problem of dealing with moral factors is to formulate a method of undergoing moral experience rather than to formulate a system of propositions about moral factors as such. Although he concedes that "moral values can only be perceived by the inner eye, which differs in each person, and is often different in the same person at different times," he argues that "there can be no doubt that experience will by itself provide a degree of objectivity to these impressions." Replicating moral experience, Clausewitz is convinced, has to be a critical component of military theory, and truths about the "sphere of mind and spirit must be rooted in experience. No theorist, and no commander, should bother himself with psychological and philosophical sophistries" (ibid.).

He then provides a formal description of the circumstances that shape the moral conditions of decision-making in war in order to explain precisely why formulating a theory of the conduct of war is difficult. The first is the critical role of moral force—that is, emotion. Clausewitz considers the most important element of emotion to be courage as a response to fear, followed by envy, generosity, pride, humility, wrath, and compassion, all of which are affected by intellectual qualities that differ widely from person to person, making the emotional factor extremely hard to measure. The second critical factor is that war consists of a series of actions and reactions by two or more adversaries whose course is inherently unpredictable. And the third is the fact that the information upon which action on both sides is based is bound to be uncertain, and the degree of uncertainty is yet another value that is difficult to quantify (ibid., II/2: 137–140). The existence of these factors means that a positive doctrine is unattainable:

> Given the nature of the subject, we must remind ourselves that it is
> simply not possible to construct a model for the art of war that can
> serve as a scaffolding on which the commander can rely for support at
> any time. Whenever he has to fall back on his innate talent, he will
> find himself outside the model and in conflict with it; no matter how
> versatile the code, the situation will always lead to the consequences
> we have already alluded to: *talent and genius operate outside the rules,
> and theory conflicts with practice.* (Ibid., II/2: 140; italics in original)

There are, Clausewitz argues, two ways out of this dilemma. First, the exercise of intelligence and judgment is much less important at lower levels of command. And something resembling positive doctrine can even play a constructive role with respect to tactical decision-making by the commander-in-chief. But at the strategic level, where the issue is not the engagement itself but its effects, and where, as a consequence, "almost all solutions must be left to imaginative intellect," Clausewitz maintains that "theory need not be a positive doctrine, a sort of manual for action," but rather a method of study (ibid., II/2: 140–141). Such "inquiry," Clausewitz wrote,

> is the most essential part of any *theory*, and which may quite appropriately claim that title. It is an analytical investigation leading to a close *acquaintance* with the subject; applied to experience—in our case, to military history—it leads to thorough *familiarity* with it. The closer it comes to that goal, the more it proceeds from the objective form of a science to the subjective form of a skill, the more effective it will prove in areas where the nature of the case admits no arbiter but talent. It will, in fact, become an active ingredient of talent. Theory will have fulfilled its main task when it is used to analyze the constituent elements of war, to distinguish precisely what at first sight seems fused, to explain in full the properties of the means employed and to show their probable effects, to define clearly the nature of the ends in view, and to illuminate all phases of warfare in a thorough critical inquiry. Theory then becomes a guide to anyone who wants to learn about war from books; it will light his way, ease his progress, train his judgment, and help him to avoid pitfalls. . . . It is meant to educate the mind of the future commander, or, more accurately, to guide him in his self-education, not to accompany him to the battlefield. (Ibid., II/2: 141; italics in original)

Clausewitz recognizes that this concept of theory can produce "principles and rules," but that these are to be thought of as a "frame of reference" rather than an instruction that lays down "precisely the path" to be taken (ibid.).

The concept of theory as a method of studying history, Clausewitz argues, "will admit the feasibility of a satisfactory theory of war" that "only needs intelligent treatment to make it conform to action, and to end the absurd difference between theory and practice that unreasonable theories have so often evoked." The subject of historical study is the "nature of ends and means" at the tactical and strategic levels. "In tactics," he explains, "the means are the fighting forces trained for combat; the end is victory." "The original means of strategy," he says, "is victory—that is, tactical success; its ends, in the final analysis, are those objects which will lead directly to peace" (ibid., II/2: 142–143).

Chapter Four

"Intelligent treatment" of the application of theory to the study of history with respect to the problem of strategy and tactics means the restriction of concern to the experience of decision-making. "A great advantage offered by this method," Clausewitz observes, "is that theory will have to remain realistic" and not "get lost in futile speculation, hairsplitting, and flights of fancy." And focusing on the moral nature of the act of decision, rather than on the taxonomy of war as a whole, means that "the range of subjects a theory must cover may be greatly simplified and the knowledge required for the conduct of war can be greatly reduced" (ibid., II/2: 144). This in fact replicates the learning conditions that produced great commanders-in-chief in the past:

> Distinguished commanders have never emerged from the ranks of the
> most erudite or scholarly officers, but have been for the most part
> men whose station in life could not have brought them a high degree
> of education. That is why anyone who thought it necessary or even
> useful to begin the education of a future general with a knowledge of
> all the details has always been scoffed at as a ridiculous pedant.
> Indeed, that method can easily be proved to be harmful: for the mind
> is formed by the knowledge and the direction of ideas it receives and
> the guidance it is given. Great things alone can make a great mind, and
> petty things will make a petty mind unless a man rejects them as
> completely alien. (Ibid., II/2: 145)

An effective capacity for the conduct of war nonetheless requires mastery of a considerable body of knowledge. "No activity of the human mind," Clausewitz states, "is possible without a certain stock of ideas; for the most part these are not innate but acquired, and constitute a man's knowledge." He warns that "genuine intellectual activity is simple and easy only in the lower ranks," and, "at the top—the position of commander-in-chief—it becomes among the most extreme to which the mind can be subjected" (ibid., II/2: 145–146). Clausewitz explains the difficulty and complexity of such intellectual demands:

> The commander-in-chief need not be a learned historian nor a
> political commentator, but he must be familiar with the higher affairs
> of state and its innate policies; he must know current issues, questions
> under consideration, the leading personalities, and be able to form
> sound judgments. He need not be an acute observer of mankind or a
> subtle analyst of human character; but he must know the character,
> the habits of thought and action, and the special virtues and defects of

the men whom he is to command. He need not know how to manage a wagon or harness a battery horse, but he must be able to gauge how long a column will take to march a given distance under various conditions. This type of knowledge cannot be forcibly produced by an apparatus of scientific formulas and mechanics; it can only be gained through a talent for judgment, and by the application of accurate judgment to the observation of man and matter. (Ibid., II/2: 146)

Acquisition of the kind of knowledge characteristic of an effective commander-in-chief requires an individual to possess a special skill, namely, the ability to assimilate the effects of experience relevant to the conduct of war from historical accounts and even from common living:

The knowledge needed by a senior commander is distinguished by the fact that it can only be attained by a special talent, through the medium of reflection, study and thought: an intellectual instinct which extracts the essence from the phenomena of life, as a bee sucks honey from a flower. In addition to study and reflection, life itself serves as a source. (Ibid.)

Clausewitz reserves description of the form of study and reflection for later chapters. But he makes it clear that a commander-in-chief's application of truths learned had to be nothing less than the "expression of his own personality." To achieve such a thing, Clausewitz explains, means that knowledge

must be so absorbed into the mind that it almost ceases to exist in a separate, objective way. In almost any other art or profession a man can work with truths he has learned from musty books, but which have no life or meaning for him. Even truths that are in constant use and are always to hand may still be externals. . . . Continual change and the need to respond to it compels the commander to carry the whole intellectual apparatus of his knowledge within him. He must always be ready to bring forth the appropriate decision. By total assimilation with his mind and life, the commander's knowledge must be transformed into a genuine capability. (Ibid., II/2: 147)

The third and fourth chapters of Book II clarify problematical points of terminology. In the third chapter, Clausewitz considers the appropriateness of the terms "art of war" and "science of war" as the names of his concept of theory. "Art," he argues, is about ability, while "science" is about knowledge. Each contains elements of the other, and thus the two terms cannot be separated entirely. But in any case, Clausewitz is convinced that neither term can be applied to his approach to the conduct of war. War, he concludes, "does not belong in the realm of arts and sciences; rather it is part of Man's social existence." "The essential difference," Clausewitz observes,

> is that war is not an exercise of the will directed at inanimate matter, as is the case with the mechanical arts, or at matter which is animate but passive and yielding, as is the case with the human mind and emotions in the fine arts. In war, the will is directed at an animate object that *reacts*. It must be obvious that the intellectual codification used in the arts and sciences is inappropriate to such activity. At the same time it is clear that continual striving after laws analogous to those appropriate to the realm of inanimate matter was bound to lead to one mistake after another. Yet it was precisely the mechanical arts that the art of war was supposed to imitate. The fine arts were impossible to imitate, since they themselves do not yet have sufficient laws and rules of their own. (Ibid., II/3: 149; italics in original)

Clausewitz does not yet give his approach a distinctive name. Instead he poses two questions, asking "whether a conflict of living forces as it develops and is resolved in war remains subject to general laws, and whether these can provide a useful guide to action." His answer is that the conduct of war, like any other subject "that does not surpass man's intellectual capacity, can be elucidated by an inquiring mind, and its internal structure can to some degree be revealed." "That alone," Clausewitz writes, "is enough to turn the concept of theory into reality" (ibid., II/3: 149–150).

In the fourth chapter, Clausewitz examines the hierarchy of terms that describes what he calls method and routine. These are law, principle, rule, regulation or direction, and method. Law, he observes, is absolute in operation, admitting no exception. Clausewitz thus rejects its use, arguing that "in the conduct of war, perception cannot be governed by laws: the complex phenomena of war are not so uniform, nor the uniform phenomena so complex, as to make laws more useful than the simple truth." The other terms, which are more flexible in character, are essential for dealing with matters at the tactical level. "Principles, rules, regulations, and methods," he writes, "are, however, indispensable concepts to or for that part of the theory of war that leads to positive doctrines; for in these doctrines the truth can express itself only in such compressed forms" (ibid., II/4: 152). At the strategic level, however, the use of method and routine "disappears completely":

> War, in its highest forms, is not *an infinite mass of minor events,* analogous despite their diversities, which can be controlled with greater or lesser effectiveness depending on the methods applied. War consists rather of *single, great decisive actions,* each of which needs to be handled individually. War is not like a field of wheat, which, without regard to the individual stalk, may be mown more or less efficiently depending on the quality of the scythe; it is like a stand of mature trees in which the axe has to be used judiciously according to the characteristics and development of each individual trunk. (Ibid., II/4: 153; italics in original)

In the fifth chapter of Book II, Clausewitz's views on the relationship between history and theory amount to a description of the historical reenactment of supreme command. He calls this procedure "critical analysis":

> The influence of theoretical truths on practical life is always exerted more through critical analysis than through doctrine. Critical analysis being the application of theoretical truths to actual events, it not only reduces the gap between the two but also accustoms the mind to these truths through their repeated application. (Ibid., II/5: 156)

Clausewitz characterizes conventional history as "the plain narrative of a historical event, which merely arranges facts one after another, and at most touches on their immediate causal links." Critical analysis, in contrast, integrates historical fact with the identification of the causes and effects of decision-making by the commander-in-chief, and the evaluation of the means—that is, the decision-making acts—employed by the commander-in-chief. The "discovery and interpretation" of historical facts is a matter of "historical research proper, and has nothing in common with theory." Critical analysis proper concerns "the tracing of effects back to their causes" and "the investigation and evaluation of means employed," the latter activity "involving praise and censure." These components, Clausewitz argues, "are the truly critical parts of historical inquiry" (ibid., II/5: 156).

Critical analysis proper is essential to productive learning from history because of gaps in the historical record that preclude complete examination of cause and effect and make it impossible to evaluate the merit of acts of decision-making. Here Clausewitz undoubtedly has in mind, among other things, the play of general friction:

> The deduction of effect from cause is often blocked by some
> insuperable extrinsic obstacle: the true causes may be quite unknown.
> Nowhere in life is this so common as in war, where the facts are
> seldom fully known and the underlying motives even less so. They
> may be intentionally concealed by those in command, or, if they
> happen to be transitory and accidental, history may not have recorded
> them at all. That is why critical narrative must usually go hand in hand
> with historical research. (Ibid.)

Critical analysis is capable of compensating for incomplete historical knowledge in some but not all cases. That is to say, it is an appropriate instrument of augmentation, but it is not supposed to constitute license for pure speculation when vital information is absent. In practice, this means that historical reenactment might be productively executed for some but not all critical decisions of a past war. Clausewitz warns that

> the disparity between cause and effect may be such that the critic is
> not justified in considering the effects as the inevitable results of
> known causes. This is bound to produce gaps—historical results that
> yield no useful lesson. All a theory demands is that investigation
> should be resolutely carried on till such a gap is reached. At that point,
> judgment has to be suspended. Serious trouble arises only when
> known facts are forcibly stretched to explain effects; for this confers
> on these facts a spurious importance. (Ibid., II/5: 156–157)

Incomplete evidence is by no means the only impediment to an accurate comprehension of past events and their meaning. Events, Clausewitz argues, are likely to have multiple causes, and the characteristics of these causes, the motives that lay behind the acts that bring them into existence, and the effects of the acts all require theoretical treatment:

> Effects in war seldom result from a single cause; there are usually
> several concurrent causes. It is therefore not enough to trace, however
> honestly and objectively, a sequence of events back to their origin:
> each identifiable cause still has to be correctly assessed. This leads to a
> closer analysis of the nature of these causes, and in this way critical
> investigation gets us into theory proper.

A critical *inquiry*—the examination of the means—poses the
question as to what are the peculiar effects of the means employed,
and whether these effects conform to the intention with which they
were used.

The particular effects of the means leads us to an investigation of
their nature—in other words, into the realm of theory again. (Ibid., II/
5: 157; italics in original)

Clausewitz is convinced that historical study that does not have the support
of sound theory will only foment confusion and controversy—that is to say,
study of history that does not take into account critical factors for which there is
no historical record, but whose effects can be given their due through reference
to theory, will be an inadequate and misleading basis for the study of command
decision. "We have seen," he writes, "both investigation of the causes and examination of the means leads to the realm of theory—that is, to the field of universal truth that cannot be inferred merely from the individual instance under
study," and that "if a usable theory does indeed exist, the inquiry can refer to its
conclusions and at that point end the investigation." In the absence of sound
theory that offers clear and generally acceptable guidelines for surmise and evaluation that facilitates simple and straightforward exposition, analysis of historical events will depend on tendentious examination of detailed but irrelevant
narrative, whose findings will most likely appear to be arbitrary. Criticism will
then fail "to reach that point at which it becomes truly instructive—when its arguments are convincing and cannot be refuted" (ibid.).

Clausewitz does not believe that the application of theory to the study of history should be mechanical. "A critic," he observes, "should never use the results of
theory as laws and standards, but only—as the soldier does—as *aids to judgment*."
This is not a difficult problem when the means and ends are "closely linked," as is
the case at the tactical level (ibid., II/5: 158; italics in original). But strategic outcomes are the product of a broad range of happenings from the tactical level to the
strategic, which, moreover, affect each other. Here again, Clausewitz is almost certainly addressing, among other things, the play of general friction. He warns:

In war, as in life generally, all parts of a whole are interconnected and
thus the effects produced, however small their cause, must influence
all subsequent military operations and modify their final outcome to
some degree, however slight. In the same way, every means must
influence even the ultimate purpose.

One can go on tracing the effects that a cause produces so long as it seems worthwhile. In the same way, a means may be evaluated, not merely with respect to its immediate end: that end itself should be appraised as a means for the next and highest one; and thus we can follow a chain of sequential objectives until we reach one that requires no justification, because its necessity is self-evident. In many cases, particularly those involving great and decisive actions, the analysis must extend to the *ultimate objective,* which is to bring about peace. (Ibid., II/5: 158–159; italics in original)

The application of theory to history is further complicated by the need to examine not only what occurred, but what could have occurred had decision-making been other than what it was. "We . . . have to consider," Clausewitz maintains, "the full extent of everything that has happened, or might have happened," which means making "a great many assumptions . . . about things that did not actually happen but seemed possible, and that, therefore, cannot be left out of account" (ibid., II/5: 159):

Critical analysis is not just an evaluation of the means actually employed, but of *all possible means*—which first have to be formulated, that is, invented. One can, after all, not condemn a method without being able to suggest a better alternative. No matter how small the range of possible combinations may be in most cases, it cannot be denied that listing those that have not been used is not a mere analysis of existing things but an achievement that cannot be performed to order since it depends on the creativity of the intellect. (Ibid., II/5: 161; italics in original)

Clausewitz says that sound theory is also essential whenever evaluating the merits of an alternative course of action regarded as superior to the one actually taken. Military writers frequently assert in hindsight that major errors could have been avoided if different actions had been taken, but they fail to demonstrate the validity of these statements through reference to sound theory. As a consequence, Clausewitz laments, the "whole literature of war" is filled with barren controversy:

The proof that we demand is needed whenever the advantage of the means suggested is not plain enough to rule out all doubts; it consists in taking each of the means and assessing and comparing the particular merits of each in relation to the objective. Once the matter has thus been reduced to simple truths, the controversy must either stop, or at least lead to new results. (Ibid., II/5: 163)

In addition to history that is augmented by theory, Clausewitz calls for the use of pure history. "In the study of means," he writes, "the critic must naturally frequently refer to military history, for in the art of war experience counts more than any amount of abstract truths" (ibid., II/5: 164). He reserves further discussion of this point for the next and final chapter of Book II.

Clausewitz believes that reenactment of a historical event can take several forms. If the critic's objective is to "distribute praise or blame, he must certainly try to put himself exactly in the position of the commander." What can be called "authentic reenactment," however, is difficult for two reasons. First, shortcomings in the historical record mean "the critic will always lack much that was present in the mind of the commander." And second, "it is even more difficult for the critic to shut off his superfluous knowledge"—that is, knowledge of the outcome of the historical event and knowledge of circumstances that were not known to the commander at the time of the battle. Clausewitz concedes that authentic reenactment can be done "well enough to suit practical purposes, but we must not forget that sometimes it is completely impossible" (ibid., II/5: 164–165).

The assessment of the rightness or wrongness of a historical act, however, is not the only objective of critical analysis through reenactment. For Clausewitz, a more valuable form is that which allows the critic to perceive the working of the commander-in-chief's genius. Such an exercise, which can be called "procedural reenactment," places as much emphasis on the accurate perception of the decision-making process as it does on the judgment of merit on the basis of outcomes:

It is . . . neither necessary nor desirable for the critic to identify himself completely with the commander. In war, as in all skills, a trained natural aptitude is called for. This virtuosity may be great or small. If it is great, it may easily be superior to that of the critic: what student would lay claim to the talent of a Frederick or a Bonaparte? Hence, unless we are to hold our peace in deference to outstanding talent, we must be allowed to profit from the wider horizons available to us. A critic should therefore not check a great commander's solution to a problem as if it were a sum in arithmetic. Rather, he must recognize with admiration the commander's success, the smooth unfolding of events, the higher workings of his genius. The essential interconnections that genius had divined, the critic has to reduce to factual knowledge.

To judge even the slightest act of talent, it is necessary for the
critic to take a more comprehensive point of view, so that he, in
possession of any number of objective reasons, reduces subjectivity to
the minimum, and so avoids judging by his own, possibly limited,
standards. (Ibid., II/5: 165–166)

Clausewitz nonetheless recognizes that a large proportion of the workings
of genius cannot be reduced to factual knowledge. For this reason, he holds that
the outcome—that is, the success or nonsuccess of operations resulting from the
commander's decisions—is germane to the perception of the action of genius,
however unquantifiable or indescribable this genius might be. What Clausewitz
seems to have in mind here is that a successful outcome invites one to resort to
theory-based surmise about decision-making dynamics:

The critic, then, having analyzed everything within the range of
human calculation and belief, will let the outcome speak for that part
whose deep, mysterious operation is never visible. The critic must
protect this unspoken result of the workings of higher laws against the
stream of uninformed opinion on the one hand, and against the gross
abuses to which it may be subjected on the other.
Success enables us to understand much that the workings of
human intelligence alone would not be able to discover. That means
that it will be useful mainly in revealing intellectual and psychological
forces and effects, because these are least subject to reliable evaluation,
and also because they are so closely involved with the will that they
may easily control it. (Ibid., II/5: 167)[6]

In the concluding paragraphs to the fifth chapter of Book II, Clausewitz
summarizes his views of procedural reenactment. In essence, he makes it clear
that proper learning has to avoid the prescriptive use of theory and instead use
theory to augment verifiable historical fact to create the basis for what is in effect
the historical reenactment of the decision-making process of the commander-
in-chief. "Critical analysis," Clausewitz explains, "is nothing but thinking that
should precede the action." Furthermore:

theory is not meant to provide [the commander-in-chief] with
positive doctrines and systems to be used as intellectual tools.
Moreover, if it is never necessary or even permissible to use scientific
guidelines in order to judge a given problem in war, if the truth never
appears in systematic form, if it is not acquired deductively but always

directly through the natural perception of the mind, then that is the way it must also be in critical analysis.

We must admit that wherever it would be too laborious to determine the facts of the situation, we must have recourse to the relevant principles established by theory. But in the same way as in war these truths are better served by a commander who has absorbed their meaning in his mind rather than one who treats them as rigid external rules, so the critic should not apply them like an external law or an algebraic formula whose relevance need not be established each time it is used. These truths should always be allowed to become self-evident, while only the more precise and complex proofs are left to theory. . . .

The complex forms of cognition should be used as little as possible, and one should never use elaborate scientific guidelines as if they were a kind of truth machine. Everything should be done through the natural workings of the mind. (Ibid., II/5: 168; italics in original)

In the sixth and final chapter of Book II, Clausewitz considers the "proper and improper uses" uses of historical examples. "Historical examples clarify everything," he writes, "and also provide the best kind of proof in the empirical sciences." That said, he warns that "historical examples are, however, seldom used to such good effect," and indeed "the use of them by theorists normally not only leaves the reader dissatisfied but even irritates his intelligence" (ibid., II/6: 170). Not all important issues in military theory, moreover, can be proved through historical examples. As Clausewitz had argued in the previous chapter, matters related to the effects of mind, temperament, and experience require the augmentation of particular historical facts with surmise based upon knowledge of the likely general dynamics of decision-making in war. In the "empirical sciences," he observes,

the theory of the art of war included, [theorists] cannot always back their conclusions with historical proofs. The sheer range to be covered would often rule this out; and, apart from that, it might be difficult to point to actual experience on every detail. . . . Theory is content to refer to experience in general to indicate the origin of the method, but not to prove it. (Ibid., II/6: 171)

Clausewitz then describes four valid methods of using history, two of which involve history as example, and two of which involve history as proof. In the former category, historical examples can be used as "an *explanation* of an idea" and "to show the *application* of an idea." In the first case, an author "may use an historical example to throw the necessary light on his idea and to ensure that the reader and the writer will remain in touch." The second case refers to the use of theory to compensate for gaps in the historical record in order to take account of the effects of mind, temperament, and experience described in Chapter 5—that is, historical reenactment. The examination of the application of an idea, Clausewitz thus explains, "gives one the opportunity of demonstrating the operation of all those minor circumstances which cannot be included in a general formulation of the idea. Indeed, this is the difference between theory and experience" (ibid.; italics in original).

Clausewitz is convinced that history unadorned by theory can be used to suggest or even demonstrate the validity of certain general propositions. Thus he argues that "one can appeal to historical fact to support a statement" and that "the detailed presentation of a historical event, and the combination of several events, make it possible to deduce a doctrine: the proof is in the evidence itself." In the two cases of using historical examples (explanation and application), "authenticity" is "not essential," and in the case of suggesting validity, the mere "simple statement of an undisputed fact" is sufficient (ibid.). But in the fourth case—proof of a general proposition that amounts to the deduction of doctrine—comprehensive analysis of verifiable historical fact is highly desirable, if not essential:

> If . . . some historical event is being presented in order to demonstrate a general truth, care must be taken that every aspect bearing on the truth at issue is fully and circumstantially developed—carefully assembled, so to speak, before the reader's eyes. To the extent that this cannot be done, the proof is weakened, and the more necessary it will be to use a number of cases to supply the evidence missing in that one. (Ibid., II/6: 171–172)

The complex authenticity required to demonstrate the validity of a general truth can only be achieved through the examination of a single case about which a great deal is known. Clausewitz is open to the use of several cases when knowledge of a single one is inadequate, but he warns that "this is clearly a dangerous expedient, and is frequently misused." He sees little value and even pernicious effect in examining an event about which information is sparse, such as a battle or campaign in the distant past. "An event that is lightly touched upon, instead of being carefully detailed," Clausewitz writes, "is like an object seen at a great distance: it is impossible to distinguish any detail, and it looks the same from every angle" (ibid., II/6: 172, 174).

The remarks that close Book II are undoubtedly self-referential. Clausewitz's personal involvement in war had been intensive and extensive. In Book II he had called for the reenactment of command decision and the use of history to prove a general proposition, the latter being, as Book VI was to show, that the defense is the stronger form of war. Both recommendations required laborious explication and contained elements that challenged conventional wisdom to such a degree that its author could not have avoided the imputation of presumption and even exposure to politically motivated reprisal. Clausewitz believed, moreover, that the future survival of his country might depend upon the adoption of his ideas. In the last two paragraphs of Book II, he thus maintains that the capacity to teach the conduct of war through historical example "would be an achievement of the utmost value," and although "it would be more than the work of a lifetime," it could nonetheless be attempted by someone who had "a thorough personal experience of war." "Anyone who feels the urge to undertake such a task," he goes on to say, "must dedicate himself for his labors as he would prepare for a pilgrimage to distant lands." Furthermore, Clausewitz warns, such a person "must spare no time or effort, fear no earthly power or rank, and rise above his own vanity or false modesty in order to tell, in accordance with the expression of the *Code Napoléon, the truth, the whole truth, and nothing but the truth*" (ibid., II/6: 174; italics in original).

Defense and Attack

Clausewitz reserves his main treatment of the question of defense to Books VI and VII, but he does raise the issue in earlier books. In the first chapter of Book I, he argues that the weaker the attacker's motive—which by definition had political/policy origins—the greater the negative effect of the relative superiority of the defense over the attack on the attacker's willingness to seek decisive action. Inaction on the part of the attacker, however, would promote commensurate behavior in the defender. "As we shall show," he declares, "defense is a stronger form of fighting than attack." "I am convinced," Clausewitz goes on to say,

> that the superiority of the defensive (if rightly understood) is very
> great, far greater than appears at first sight. It is this which explains

without any inconsistency most periods of inaction that occur in war. The weaker the motives for action, the more will they be overlaid and neutralized by this disparity between attack and defense, and the more frequently will action be suspended—as indeed experience shows. (Ibid., I/1: 84)

In the real world, Clausewitz writes, the attacker cannot achieve a decision with "a single, short blow." In the time that elapses between the initiation of attack and the occurrence of major battle, the defender can mobilize or deploy additional regular troops augmented by a supportive population; exploit, through skillful retreat, the effects of topography, distance, and fixed defenses; and receive the support of allies. "Even when great strength has been expended on the first decision," Clausewitz maintains, "and the balance has been badly upset, equilibrium can be restored" (ibid., I/1: 79).

In the second chapter of Book I, he observes that an attacker not only must destroy the fighting forces of the defender, but must occupy the country and break the defender's will to resist in order to forestall popular insurrection. He then examines the conditions of defense against an attack by a much stronger enemy. Under such circumstances, Clausewitz states, the proper objective of the defense must not be to disarm the enemy, but rather "wearing down" the invader, which means "using the duration of the war to bring about a gradual exhaustion of his physical and moral resistance" (ibid., I/2: 90–91, 93). Adopting the perspective of the defense, Clausewitz then argues that

if we intend to hold out longer than our opponent we must be content with the smallest possible objects, for obviously a major object requires more effort than a minor one. The minimum object is *pure self-defense;* in other words, fighting without a positive purpose. With such a policy our relative strength will be at its height, and thus the prospects for a favorable outcome will be the greatest. (Ibid., I/2: 99; italics in original)

Clausewitz believes that it is justifiable for the defense to act to preserve its forces without combat if a large imbalance in strength makes such a course necessary. But he also emphasizes the fundamental importance of the destruction of the enemy's forces through fighting. These are not necessarily mutually exclusive forms of behavior. Clausewitz resolves the apparent contradiction by explaining that the preservation of one's own forces and the exhaustion of those of the enemy can also be the preliminary to acting with a positive purpose, that is, attempting to destroy the enemy forces through fighting. In this case, the action to preserve one's own forces

> is transposed into waiting for the decisive moment. This usually means that *action is postponed* in time and space to the extent that space and circumstances permit. If the time arrives when further waiting would bring excessive disadvantages, then the benefit of the negative policy has been exhausted. The destruction of the enemy— an aim that has until then been postponed but not displaced by another consideration—now reemerges. (Ibid.; italics in original)

In the third chapter of Book I, Clausewitz describes the salient characteristics of an effective commander-in-chief, upon which the fortunes of either offensive or defensive action will depend. For Clausewitz, such "military genius," as has been explained previously, is a matter of mind, temperament, and experience. Although he did not say so at this time, these challenges are self-evidently pronounced for the leader of the attacking force, but likely to be attenuated for the general in charge of the defense—in the latter case, especially so when action is avoided entirely in order to preserve forces. Continuing to set the stage for an explicit discussion of the psychological difficulties inherent to the attack and their favorable implications for the defense, which he provides in Books VI and VII,[7] Clausewitz describes the major factors of friction—and the antidote to friction, namely experience—in the next five chapters of Book I. An unstated though obvious implication is that although this factor could not be available to an attacker at the beginning of a conflict that started after many years of peace, it might be possessed by the defender in some measure at the time of counterattack following a protracted period of armed resistance.

In Books II, III, IV, and V, Clausewitz makes observations that support his views on the superiority of the defense over the offense. In the fifth chapter of Book II, he alludes to the strategy of protracted defensive war when he discusses the power of Napoleon's reputation to inspire fear in his opponents and bring them to terms. In the Italian campaign of 1797, Clausewitz observes, the Austrians sought peace at Campo Formio—in spite of the fact that their military forces, though outnumbered, were still intact—because "the secret of the effectiveness of resisting to the last had not yet been discovered" (ibid., II/5: 161). In Book III, entitled "Strategy in General," Clausewitz deals with three subjects that favor the proposition that defense is the stronger form of war. These are the critical importance of the moral element in strategy (ibid., III/3: 184–185), the enormous psychological difficulty of being able to use all forces simultaneously (which Clausewitz believes is important for the attacker) (ibid., III/12: 209), and the suspension of action in war (ibid., III/16: 216–219). In Book IV, "The Engagement," he argues that the attacker needs a great battle to achieve decision in the event of resistance by the defender, but that mustering the will to accomplish this is extremely difficult (ibid., IV/11: 259). Moreover, he contends that the pursuit and destruction of an enemy army after a great battle, which he considers to be essential for strategic success, poses especially difficult challenges of will for the victorious commander-in-chief (ibid., IV/12: 263–270). In Book V, "Military Forces," he maintains that even a relatively weak defender can resist effectively (ibid., V/3: 283), that swift offensive action through rapid marching is exhausting and can seriously debilitate an attacking army (ibid., V/12: 322), and that, in the event of a delayed decision, both inadequate billeting and logistics are more likely to afflict the attacker than the defender during the period of suspended hostilities (ibid., V/14: 339–340).

"A whole range of propositions," Clausewitz maintained in his "Unfinished Note," "can be demonstrated without difficulty." Among these was the assertion that "the defense is the stronger form of fighting with the negative purpose, attack the weaker form with the positive purpose" (ibid., 71).[8] He presents his proof of this argument in Books VI and VII of *On War*. As specified in Chapter 6 of Book II, his exposition is heavily based on historical examples, and "every aspect bearing on the truth at issue" is "fully and circumstantially developed." As a consequence, Book VI is more than twice the length of the next longest Book (Book V) and more than triple the length of most of the others (Books I, II, III, and VII). Book VII adds even greater bulk to the exposition of his views on the superiority of the defense over the offense.

Clausewitz divides Book VI into three main sections. He states that in the first eight chapters, he "surveyed as well as delimited the whole field of defense" (ibid., VI/8: 385). He maintains that in Chapters 9 through 26 he covered "the most important methods of defense," and that in Chapters 27 through 30 he examined "the defense of a theater of war as a subject in itself" and looks "for the thread that ties together all the subjects discussed"(ibid., VI/27: 484). The first and last sections are each approximately thirty pages and constitute the introduction and conclusion. The middle section is roughly 100 pages long—or twice the length of most of the other books of *On War*. In Book VI, Clausewitz presents a long and carefully rendered proof of a major proposition. It cannot be dismissed out of hand as a sketch, defective trial run, or mere compendium of obsolete technical observations.

Clausewitz's basic definition of defense, which he supplies in the first chapter of Book VI, has two elements: "the parrying of a blow" and "awaiting the blow." Awaiting the blow is critical to a valid definition because "it is the only test by which defense can be distinguished from attack in war" (ibid., VI/1: 357). Clausewitz argues that the great advantage of defense over offense is

> the fact that time which is allowed to pass unused accumulates to the credit of the defender. He reaps where he did not sow. Any omission of attack—whether from bad judgment, fear, or indolence—accrues to the defenders' benefit. (Ibid.)

That said, Clausewitz makes it clear that parrying and awaiting do not rule out offensive action:

> If defense is the stronger form of war, yet has a negative object, it follows that it should be used only so long as weakness compels, and be abandoned as soon as we are strong enough to pursue a positive object. When one has used defensive measures successfully, a more favorable balance of strength is usually created; thus, the natural course in war is to begin defensively and end by attacking. It would therefore contradict the very idea of war to regard defense as its final purpose, just as it would to regard the passive nature of defense not only as inherent in the whole but also in all its parts. In other words, a war in which victories were used only defensively without the intention of counterattacking would be as absurd as a battle in which the principle of absolute defense—passivity, that is—were to dictate every action. (Ibid., VI/1: 358)

In Chapters 2 through 4, Clausewitz examines the relative merits of the offense and defense. In the second chapter, he argues that the replacement of passive cordon defense by mobile defense in depth during the Wars of the French Revolution and Empire had shifted the balance of defense and offense in favor of the former (ibid., VI/2: 362). In the third chapter, he first examines the weaknesses of the offense by discounting the significance of offensive strategic surprise and initiative, ruling out the offensive use of strategic concentric attack, observing that offensive strategic action creates vulnerabilities that can be exploited by defensive counterattack, and noting that moral forces favoring the attacker do not come into play until after the decisive blow has been struck. Clausewitz then enumerates the strengths of the defense: the ability to gain strength in retreat because of the support of fortresses, the shortening of supply lines, and the action of militias and armed civilians (ibid., VI/3: 363–366). In the fourth chapter, he posits that maneuver and operational depth can enable a defender to exploit the advantages of interior lines and greater concentration (ibid., VI/4: 368).

In the fifth and sixth chapters of Book VI, Clausewitz focuses his analysis on the specific characteristics of defense. He begins by stating that it is action by the defender, rather than by the attacker, that determines that war should occur, which suggests that from the outset of war, it is the defender that holds the initiative:

> War serves the purpose of the defense more than that of the aggressor. It is only aggression that calls forth defense, and war along with it. The Aggressor is always peace-loving (as Bonaparte always claimed to be); he would prefer to take over our country unopposed. To prevent his doing so one must be willing to make war and be prepared for it. In other words it is the weak, those likely to need defense, who should always be armed in order not to be overwhelmed. (Ibid., VI/5: 370)

Clausewitz also declares his views on the crucial importance of counterattack:

> Once the defender has gained an important advantage, defense as such has done its work. While he is enjoying this advantage, he must strike back, or he will court destruction. Prudence bids him strike while the iron is hot and use the advantage to prevent a second onslaught. . . . This transition to the counterattack must be accepted as a tendency inherent in defense—indeed, as one of its essential features. Wherever a victory achieved by the defensive form is not turned to military account, where, so to speak, it is allowed to wither away unused, a serious mistake is being made. A sudden powerful transition to the offensive—the flashing sword of vengeance—is the greatest moment for the defense. (Ibid.)

Defense as it should be, he concludes, calls for the following:

> All means are prepared to the utmost; the army is fit for war and familiar with it; the general will let the enemy come on, not from

confused indecision and fear, but by his own choice, coolly and deliberately; fortresses are undaunted by the prospect of a siege, and finally a stout-hearted populace is no more afraid of the enemy than he of it. Thus constituted, defense will no longer cut so sorry a figure when compared to attack, and the latter will no longer look so easy and infallible as it does in the gloomy imagination of those who see courage, determination, and movement in attack alone, and in defense only impotence and paralysis. (Ibid., VI/5: 371)

Clausewitz provides a systematic reprise of his examination of the character of the defense by enumerating its major resources: militia, fortresses, and the favorable disposition of a country's inhabitants to its government, its armed civilians, and its allies. With respect to the last, Clausewitz states that "as a rule the defender can count on outside assistance more than can the attacker; and the more his survival matters to the rest—that is, the sounder and more vigorous his political and military condition—the more certain he can be of their help" (ibid., VI/6: 376).

In the seventh chapter of Book VI, Clausewitz returns to the matter of the interaction of the offense and defense and amplifies views given in Chapter 5. "The idea of war," he maintains, "originates with the defense, which does have fighting as its immediate object, since fighting and parrying obviously amount to the same thing. . . . It is the defender, who not only concentrates his forces but disposes them in readiness for action, who first commits an act that really fits the concept of war" (ibid., VI/7: 377).

In Chapter 8, which concludes his introduction to the subject of defense, Clausewitz articulates his position with clarity and force. "The essence of defense," he begins, "lies in parrying the attack." Defense so defined has two distinct parts, namely, "waiting and acting." As he stated previously, Clausewitz considers waiting to be a concept of major importance. "Waiting," he observes, "is such a fundamental feature of all warfare that war is hardly conceivable without it, hence we shall often have occasion to revert to it by pointing out its effect in the dynamic play of forces. If waiting is "the salient feature and chief advantage" of defense, action is its necessary consummation (ibid., VI/8: 379). He thus maintains that

waiting and acting—the latter always being a riposte and therefore a reaction—are both essential parts of defense. Without the former, it would not be defense, without the latter, it would not be war. This conception has already led us to argue that *defense is simply the stronger form of war, the one that makes the enemy's defeat more certain.* (Ibid., VI/8: 380; italics in original)

Clausewitz's view of the matter rules out both preemption and preventive war. "We must insist," he declares, "that the idea of retaliation is fundamental to all defense." "Since defense is tied to the idea of waiting," he then goes on to say,

> the aim of defeating the enemy will be valid only on the condition that there is an attack. If no attack is forthcoming, it is understood that the defense will be content to hold its own. . . . The defense will be able to reap the benefits of the stronger form of war only if it is willing to be satisfied with this more modest goal. (Ibid.)

Clausewitz gives explicit instructions as to the allowable range of actions that can be considered to be legitimately defensive. The four options are, in ascending order of defensiveness, armed resistance triggered by the enemy's entry into the theater of operations; armed resistance in response to offensive enemy movement near the national frontier; armed resistance at or behind the national frontier in response to invasion; and armed resistance from the interior of the country. Although any of the actions listed satisfies his basic definition of what constitutes a defense, Clausewitz makes it clear that defensive advantage increases from the first option to the fourth. That is, he explains, "with each successive stage of defense the defender's predominance or, more accurately, his counterweight will increase, and so in consequence will the strength of his reaction." In the first three cases of defensive action, attacker successes that do not preclude the exercise of the fourth are by definition indecisive. In the "first three stages of defense (in other words, those taking place at the border), Clausewitz observes, "*the very lack of a decision constitutes a success for the defense*" (ibid., VI/8: 380–383; italics in original).

Clausewitz divides the four cases into two forms of action. The first form applies to the first three cases, in which the period of waiting is minimal to short. Defender action during this time is directed toward the destruction of the attacker through decisive battle. The second form involves protracted waiting. Defender action—that is, inaction—is directed toward the preservation of its own military forces and the weakening of the attacker through difficulties of supply, the diversion of forces to control hostile rear areas, and the debilitating effects of weather and season. Thus, Clausewitz argues, "two decisions, and therefore two kinds of reaction, are possible on the defending side, depending upon whether the attacker is to *perish by the sword* or *by his own exertions*" (ibid., VI/8: 384; italics in original).

Clausewitz states that the "advantages of waiting" are "completely exhausted" "when the attacker has gained ground," and as a consequence the military strength of the defender begins to decline. He notes that these conditions define "the point of culmination"—that is, the moment at which "the defender must make up his mind and act" (ibid., VI/8: 383).[9] But he makes it clear that action by the defender will not inevitably precipitate a major engagement:

> In our discussion, we have always assumed decision to occur in the form of battle, but that is not necessarily so. We can think of any number of engagements by smaller forces that may lead to a change in fortune, either because they really end in bloodshed, or because the probability of their consequences necessitate the enemy's retreat. (Ibid., VI/8: 384)

In these circumstances, it is not just fighting as such, but the threat of fighting, which converts the attacker's logistical privation into strategic overextension. Clausewitz explains that "when the enemy has become the victim of the difficult conditions of the advance and has been weakened and reduced by hunger, by sickness and the need to detach troops, it is really the fear of our fighting forces alone that makes him turn about and abandon all he has gained" (ibid., VI/8: 384).

Clausewitz has much to say about the importance of the idea as opposed to the actuality of major combat:

> If an attacker finds the enemy in a strong position that he thinks he cannot take, or on the far side of a river that he believes to be impassable, or even if he fears he will jeopardize his food supply by advancing further, it is still only the force of the defender's arms which produces these results. What actually halts the aggressor's action is the fear of defeat by the defender's forces, either in major engagements or at particularly important points; but he is not likely to concede this, at least not openly.
>
> One may admit that even where the decision has been bloodless, it was determined in the last analysis by engagements that did not take place but *had merely been offered*. (Ibid., VI/8: 386; italics in original)

Clausewitz absolutely rejects the notion that attacker inaction is attributable primarily to consideration of physical circumstances, as is so often believed to be the case:

> When we look at the history of war and find a large number of
> campaigns in which the attacker broke off his offensive without having
> fought a decisive battle, consequently where strategic combinations
> appear effective, we might believe that such combinations have at least
> great inherent power, and that they would normally decide the
> outcome on their own whenever one did not need to assume a decisive
> superiority of the offensive in tactical situations. Our answer here must
> be that this assumption, too, is erroneous in situations that arise in the
> theater of operations and are therefore part of war itself. The reason for
> the ineffectiveness of most attacks lies in the general, the political
> conditions of war. (Ibid., VI/8: 386-387)

He is convinced that politics/policy has a greater negative effect on the strategy of the attacker than it does on the defender. It is the "political conditions of war" just mentioned, he insists, that

> have transformed most wars into mongrel affairs, in which the
> original hostilities have to twist and turn among conflicting interests
> to such a degree that they emerge very much attenuated. This is
> bound to affect the offensive, the side of positive action, with
> particular strength. It is not surprising, therefore, that one can stop
> such a breathless, hectic attack by the mere flick of a finger. Where
> resolution is so faint and paralyzed by a multitude of considerations
> that it has almost ceased to exist, a mere show of resistance will often
> suffice. (Ibid., VI/8: 387)

In the case of offensive (real) absolute war, Clausewitz does not provide an explicit explanation for why the attacker is more heavily affected by political/policy considerations than the defender. Presumably, it is that no matter how determined the attacker, no political/policy motive for national aggrandizement through offensive action can be equal to or stronger than the motive of national survival on the part of a defender determined to preserve its sovereignty.

If for any reason the will of the defender is greater than that of the attacker, with the result that hostilities are protracted, the latter is bound to suffer discouraging losses that will cause it to change its politics/policy objectives in ways that

favor the defender. Clausewitz believes that this is not generally understood because the governments of attacking powers that are caught in the predicament
of an arrested offensive rarely, if ever, explain their actual reasons for coming to
terms short of their original goals:

> The counterweights that weaken the elemental force of war, and
> particularly the attack, are primarily located in the political relations
> and intentions of the government, which are concealed from the rest
> of the world, the people at home, the army, and in some cases even
> from the commander. . . . If military history is read with this kind of
> skepticism, a vast amount of verbiage concerning attack and defense
> will collapse, and the simple conceptualization we have offered will
> automatically emerge. We believe that it is valid for the whole field of
> defense, and that only if we cling to it firmly can the welter of events
> be clearly understood and mastered. (Ibid., VI/8: 388)

In closing, Clausewitz indicates that he considers the material presented in
Chapter 8 to be of the first importance. "We should like to add," he writes in his
final sentence, "that this chapter, more than any other of our work, shows that
our aim is not to provide new principles and methods of conducting war; rather,
we are concerned with examining the essential contents of what has long existed, and to trace it back to its basic elements" (ibid., VI/8: 389).

The main body of Book VI is divided into three subsections. In Chapters 9
through 21, Clausewitz examines the physical dimensions of defensive action,
covering defensive battle, fortresses, various kinds of defensive positions, and defense with respect to the major forms of terrain. In these discussions, he argues
against dependence upon fixed defenses such as fortresses and terrain features
and in favor of action that maximizes the effects of freedom of maneuver in general, and counterattack in particular. In Chapters 22 to 24, Clausewitz criticizes
and dismisses certain standard concepts of defensive action that he regards as
weak. In Chapters 25 and 26, he identifies two courses of action—retreat into the
interior of the country and armed resistance by the populace—as potentially capable of producing major effects. These are, from the spatial and sociopsycholog-
ical points of view, the ultimate forms of defense in depth. Given the significance
of this matter in Clausewitz's conception of the superiority of the defense over
the offense, his specific views on these subjects deserve separate consideration.

Clausewitz bases his theoretical conclusions mainly on the Russian campaign of 1812. He observes that "voluntary withdrawal to the interior of the

country . . . destroys the enemy not so much by the sword as by his own exertions." "Debilitation in the course of an advance is increased," he adds, "if the defender is undefeated and retreats voluntarily with his fighting forces intact and alert, while by means of a steady, calculated resistance he makes the attacker pay in blood for every foot of progress." If the defense avoids a major military defeat, the attacker will not only be weakened substantially in the course of his advance but will be exposed to powerful counterattack, and the effects of this counterattack will be magnified by his isolation deep in hostile territory. Clausewitz discounts the significance of the forfeiture of human and material resources occasioned by retreat, observing that "it cannot be the object of defense to protect the country from losses; the object must be a favorable peace." He has serious concerns, however, about the negative psychological effects of large-scale withdrawal, which he recognized could demoralize both the army and the general population, and thereby weaken or even collapse the defensive effort (ibid., VI/25: 469–471).

In Chapter 26, Clausewitz examines the role of popular insurrection in support of the war effort of a national government fighting a defensive campaign. This practice, he notes, is a phenomenon that had been brought into being in the "civilized parts of Europe" by the advent in the nineteenth century of offensive action aimed at the destruction of national sovereignty—namely (real) absolute war. Although obviously informed by knowledge of the history of guerrilla warfare in Spain during the Napoleonic period, Clausewitz's discussion of People's War is entirely theoretical. "Any nation that uses it intelligently," he asserts, "will, as a rule, gain some superiority over those who disdain its use" (ibid., VI/26: 479). The effect of People's War, Clausewitz observes

> is like that of the process of evaporation: it depends on how much surface is exposed. The greater the surface and the area of contact between it and the enemy forces, the thinner the latter have to be spread, the greater the effect of a general uprising. Like smoldering embers, it consumes the basic foundations of the enemy forces. (Ibid., VI/26: 480)

Because of the considerable potential effectiveness of People's War, Clausewitz believes it has to be taken into strategic account, especially in the event of catastrophic military defeat:

> A government must never assume that its country's fate, its whole existence, hangs on the outcome of a single battle, no matter how

decisive. Even after a defeat, there is always the possibility that a turn of fortune can be brought about by developing new sources of internal strength or through the natural decimation all offensives suffer in the long run or by means of help from abroad. There will always be time enough to die; like a drowning man who will clutch instinctively at a straw, it is the natural law of the moral world that a nation that finds itself on the brink of an abyss will try to save itself by any means. (Ibid., VI/26: 483)

Clausewitz equates an unwillingness to resort to People's War with moral weakness, and a government that fails to use it when required, he says, is derelict in its duty. There is no mistaking the intended target of this barb, which has to be the behavior of King Frederick William III in 1806 and 1812:

No matter how small and weak a state may be in comparison with its enemy, it must not forego these last efforts, or one would conclude that its soul is dead. The possibility of avoiding total ruin by paying a high price for peace should not be ruled out, but even this intention will not, in turn, eliminate the usefulness of new measures of defense. They will not make the peace more difficult and onerous, but easier and better. They are even more desirable where help can be expected from other states that have an interest in our survival. A government that after having lost a major battle, is only interested in letting its people go back to sleep in peace as soon as possible, and, overwhelmed by feelings of failure and disappointment, lacks the courage and desire to put forth a final effort, is, because of its weakness, involved in a major inconsistency in any case. It shows that it did not deserve to win, and, possibly for that very reason was unable to. (Ibid.)

In the last chapters of Book VI, which are devoted to the defense of a theater of operations, Clausewitz advances four major propositions. In Chapter 27, he argues that for the defender, preservation of the army is in general more important than preservation of territory (ibid., VI/27: 485). In Chapter 28, he maintains that territory that has been abandoned as the prelude to counterattack is no less defended than if it had been contested (ibid., VI/28: 488). In Chapter 29, Clausewitz states that continuous, vigorous resistance by the defender's regular forces, combined with the negative effects of other factors previously described, will in most cases be sufficient to bring about peace that offers the attacker no more than a "modest advantage" (ibid., VI/29: 500). And finally, in Chapter 30, Clausewitz observes that when the political/policy motivation of the attacker is weak, and his actions feeble, the defender's reaction is likely to be similar, producing a situation in which no great battle will occur because neither side seeks a decision. That said, he warns that either side can at any time choose to seek a decision through more vigorous action, a danger that always has to be taken into account (ibid., VI/30: 513, 517).

In Book VII, Clausewitz amplifies his major arguments on the superiority of the defense over the offense, devoting most of his attention to the weaknesses of the attack. In the first chapter, he argues that although the strengths of the defense "may not be insurmountable, the cost of surmounting them may be disproportionate" (ibid., VII/1: 523). In the second chapter, he observes that "an attack cannot be completed in a single steady movement: periods of rest are needed, during which the attack is neutralized, and defense takes over automatically"; in addition, the attacker's rear, which is vital to its existence, requires protection. Thus he maintains that "the superiority of strategic defense arises partly from the fact that the attack itself cannot exist without some measure of defense— and defense of a much less effective kind." Moreover, "it is these very moments of weak defense during an offensive that the positive activity of the offensive principle *in defense* seeks to exploit" (ibid., VII/2: 524; italics in original).

In Chapters 3 through 5, Clausewitz examines the inherent tendency of the attack to falter. In Chapter 3, he notes that attacks in practice "often turn into defensive action" (ibid., VII/3: 526). In Chapter 4, he describes why attacks tend to diminish in strength as they progress when the objective is the occupation of the other country (ibid., VII/4: 527). He notes in Chapter 5 that most strategic attacks "lead up to the point where their remaining strength is just enough to maintain a defense and wait for peace." Moving beyond that point, which he calls the "culminating point of the attack," exposes the attacker to a counterattack that "follows with a force that is usually much stronger than that of the original attack." Determining proximity to the culminating point of attack on the part of the attacker is extremely difficult and often "entirely a matter of the imagination" (ibid., VII/5: 528). He devotes Chapters 6 through 21 to technical points. Insofar as the superiority of the defense over the attack is concerned, in Chapters 6, 8, 9, and 10 Clausewitz argues that while decisive battle is essential to the attacker, it is extremely difficult to achieve if the defender is in a good defensive position or unwilling to stand.[10]

Chapter 22, which concludes Book VII, was originally a separate essay on the "culminating point of victory." It extends and amplifies the discussion presented in Chapter 5 (ibid., VII/5: 528n1). At the outset, Clausewitz declares:

> It is clear, therefore, that a defense that is undertaken in the framework of an offensive is weakened in all its key elements. It will thus no longer possess the superiority which basically belongs to it.
>
> Just as no defensive campaign consists simply of defensive elements, so no offensive campaign consists purely of offensive ones. Apart from the short intervals in every campaign during which both sides are on the defensive, every attack which does not lead to peace must necessarily end up as a defense.
>
> It is thus defense that weakens the attack. Far from being idle sophistry, we consider it to be the greatest disadvantage of the attack that one is eventually left in a most awkward position. (Ibid., VII/22: 572)

Clausewitz then maintains that the psychological challenges of the attack are so great as to weaken the resolve of all but the most determined commanders or confound the efforts of the overly adventurous:

> In reviewing the whole array of factors a general must weigh before making his decision, we must remember that he can gauge the direction and value of the most important ones only by considering numerous other possibilities—some immediate, some remote. He must *guess*, so to speak: guess whether the first shock of battle will steel the enemy's resolve and stiffen his resistance, or whether, like a Bologna flask, it will shatter as soon as its surface is scratched; guess the extent of debilitation and paralysis that the drying up of particular sources of supply and the severing of certain lines of communication will cause in the enemy; guess whether the burning pain of the injury he has been dealt will make the enemy collapse with exhaustion or, like a wounded bull, arouse his rage; guess whether the other powers will be frightened or indignant, and whether and which political alliances will be dissolved or formed. When we realize that he must hit upon all this and much more by means of his discreet judgment, as a marksman hits a target, we must admit that such an accomplishment of the human mind is no small achievement. Thousands of wrong turns running in all directions tempt his perception; and if the range, confusion and complexity of the issues are not enough to overwhelm him, the dangers and responsibilities may.

> This is why the great majority of generals will prefer to stop well
> short of their objective rather than risk approaching it too closely, and
> why those with high courage and an enterprising spirit will often
> overshoot it and so fail to attain their purpose. (Ibid., VII/22: 573;
> italics in original)

In Book VIII, Clausewitz considers the specific dynamics of planning a war and
campaign. In the process, he also reprises and in certain important respects clarifies his main arguments pertaining to absolute war and genius, history and theory, and defense and attack.

Clausewitz reiterates that a theory of war has to cover all cases of war, from
the least to the most intense. Although the Wars of the French Revolution and
Empire had made it clear that (real) absolute war could occur, Clausewitz warns
that wars of the future might not take this form. A viable general theory of war
could not, therefore, be applicable only to those conflicts involving the maximization of violence:

> We must, therefore, be prepared to develop our concept of war as it
> ought to be fought, not on the basis of its pure definition, but by
> leaving room for every sort of extraneous matter. We must allow for
> natural inertia, for all the friction of its parts, for all the inconsistency,
> imprecision, and timidity of man; and finally we must face the fact
> that war and its forms result from ideas, emotions, and conditions
> prevailing at the time—and to be quite honest we must admit that this
> was the case even when war assumed its absolute state under
> Bonaparte. (Ibid., VIII/2: 580)

Clausewitz believes that because (real) absolute war and its demands are always
a possibility, it has to be taken into account by the commander-in-chief of an
army even in the case of (less than absolute) real war. Theory, he warns,

> has the duty to give priority to the absolute form of war and to make
> that form a general point of reference, so that he who wants to learn
> from theory becomes accustomed to keeping that point in view
> constantly, to measuring all his hopes and fears by it, and to
> approximating it *when he can* or *when he must*. (Ibid., VIII/2: 581;
> italics in original)

This is to say that because the potential for (real) absolute war is contained within [less than absolute] real war, the two forms are conjoined rather than distinct taxonomic categories until after the conflict has ended, at which time the occurrence or nonoccurrence of escalation in violence has been established as fact. This capacity of war to change its fundamental form depending upon circumstances is probably what Clausewitz is referring to when he says "war is more than a true chameleon that slightly adapts its characteristics to a given case" (ibid., I/1: 89).

As a way of defining the nature of genius, Clausewitz describes the complexity of the task of taking account of political factors when planning a war:

> To discover how much of our resources must be mobilized for war, we must first examine our own political aim and that of the enemy. We must gauge the strength and situation of the opposing state. We must gauge the character and abilities of its government and people and do the same in regard to our own. Finally, we must evaluate the political sympathies of other states and the effect the war may have on them. To assess these things in all their ramifications and diversity is plainly a colossal task. Rapid and correct appraisal of them clearly calls for the intuition of a genius; to master all this complex mass by sheer methodical examination is obviously impossible. Bonaparte was quite right when he said that Newton himself would quail before the algebraic problems it could pose. (Ibid., VIII/3: 585–586)

Clausewitz recognizes that strategic decision-making before the outbreak of war will depend upon "the qualities of mind and character of the men making the decision—of the rulers, statesmen, and commanders, whether these roles are united in a single individual or not" (ibid., VIII/3: 586). He is nonetheless convinced that during a war, the main executor of decision—and thus the venue for the operation of genius—is the commander-in-chief:

> If war is to be fully consonant with political objectives, and policy suited to the means available for war, then unless statesman and soldier are combined in one person, the only sound expedient is to make the commander-in-chief a member of the cabinet, so that the cabinet can share in the major aspect of his activities. . . . What is highly dangerous is to let any soldier but the commander-in-chief exert an influence in the cabinet. It very seldom leads to sound vigorous action. (Ibid., VIII/6: 608–609)

Clausewitz warns against the improper use of theory and alludes to his alternative approach of using history and theory in combination as the basis of reenactment of the decision-making genius of the commander-in-chief. He confesses that he approaches the consideration of war planning in general terms with "some diffidence" (ibid., VIII/1: 577). "We are overcome with the fear," he writes,

> that we shall be irresistibly dragged down to a state of dreary pedantry, and grub around in the underworld of ponderous concepts where no great commander, with his effortless *coup d'oeil,* was ever seen. If that were the best that theoretical studies could produce it would be better never to have attempted them in the first place. Men of genuine talent would despise them and they would quickly be forgotten. When all is said and done, it really is the commander's *coup d'oeil,* his ability to see things simply, to identify the whole business of war completely with himself, that is the essence of good generalship. Only if the mind works in this comprehensive fashion can it achieve the freedom it needs to dominate events and not be dominated by them. (Ibid., VIII/1: 578)

The purpose of theory, Clausewitz goes on to explain, is improved perception of the particular, not prescription:

> Theory should cast a steady light on all phenomena so that we can more easily recognize and eliminate the weeds that always spring from ignorance; it should show how one thing is related to another, and keep the important and the unimportant separate. . . . Theory cannot equip the mind with formulas for solving problems nor can it mark the narrow path on which the sole solution is supposed to lie by planting a hedge of principles on either side. But it can give the mind insight into the great mass of phenomena and of their relationships, then leave it free to rise into the higher realms of action. (Ibid.)

Clausewitz does not rule out recognition and application of a military proposition that is validated by reasonable examination of the historical evidence. "If concepts combine of their own accord to form that nucleus of truth we call a principle," he observes, and "if they spontaneously compose a pattern that becomes a rule, it is the task of the theorist to make this clear" (ibid.). That the defense is the stronger form of war in both cases—that is, of (real) absolute and (less than absolute) real war—Clausewitz is convinced, is demonstrated by history. With respect to the latter, he examines the difficulties of consolidating offensive gains in the great wars of the mid-eighteenth century (ibid., VIII/7: 611–612). His prime example of the former is the Russian campaign of 1812:

> If we wish to learn from history, we must realize that what happened once can happen again; and anyone with judgment in these matters will agree that the chain of great events that followed the march on Moscow was no mere succession of accidents. To be sure, had the Russians been able to put up any kind of defense of their frontiers, the star of France would probably have waned, and luck would probably have deserted her; but certainly not on that colossal and decisive scale. It was a vast success; and it cost the Russians a price in blood and perils that for any other country would have been higher still, and which most could not have paid at all. (Ibid., VIII/8: 616)

Addressing the issue of the relative merits of the defense and the attack, Clausewitz argues that "the aim of the defense must embody the idea of waiting—which is after all its leading feature." He admits that waiting alone can sometimes satisfy the requirements of defense. "Certainly," he writes, "the exhaustion, or, to be accurate, the fatigue of the stronger has often brought about peace" (ibid., VIII/8: 613). Nevertheless, Clausewitz states,

> a defender must always seek to change over to the attack as soon as he has gained the benefit of the defense. So it follows that among the aims of such an attack, which is to be regarded as the real aim of the defense, however significant or insignificant this may be, the defeat of the enemy could be included. (Ibid., VIII/8: 600)

On the issue of the effects of protracted hostilities, Clausewitz argues, from the point of view of the attacker, that "no conquest can be carried out too quickly, and that to spread it over a *longer period* than the minimum needed to complete it *makes it not less difficult, but more*" (ibid., VIII/4: 598; italics in original). Furthermore,

> any kind of interruption, pause, or suspension of activity is inconsistent with the nature of offensive war. When they are unavoidable, they must be regarded as necessary evils, which make success not more but less certain. Indeed, if we are to keep strictly to the truth, when weakness does compel us to halt, a second run at the objective normally becomes impossible; and if it does turn out to be

possible it shows that there was no need for a halt at all. When an objective was beyond one's strength in the first place, it will always remain so. (Ibid., VIII/4: 600)

There can be little doubt that Clausewitz believes that the costs of achieving large political objectives—that is to say, those that threaten the national existence of the defender and thus require great and sustained military exertion—are almost always more than even the greatest powers can bear:

> The conditions for defeating an enemy presuppose great physical or moral superiority or else an extremely enterprising spirit, an inclination for serious risks. When neither of these is present, the object of military activity can only be one of two kinds: seizing a small or larger piece of enemy territory, or holding one's own until things take a better turn. (Ibid., VIII/5: 601)

In most wars, Clausewitz thus concludes, "the future seems to promise nothing definite to either side and hence affords no grounds for a decision" (ibid., VIII/5: 602). The result is that "interaction, the effort to outdo the enemy, the violent and compulsive course of war, all stagnate for lack of real incentive. Neither side makes more than minimal moves, and neither feels itself seriously threatened" (ibid., VIII/6: 604). And as he had explained previously, any delay in action and thus in a decision favors the defense.

Clausewitz had made clear in Book VI that decisions are not deferred because of physical military factors but because the attacker lacks the will to act. The attacker's lack of will, in turn, is the product of political considerations. It is for this reason that Clausewitz declares that a unified view of war is one that recognizes that *the concept that war is only a branch of political activity; that it is in no sense autonomous.* He goes on to explain that political/policy considerations not only govern the decision to initiate hostilities but are "influential in the planning of war, of the campaign, and often even of the battle" (ibid., VIII/6: 605–606; italics in original): He thus maintains that

> the probable character and general shape of any war should mainly be assessed in the light of political factors and conditions—and that war should often (indeed today one might say normally) be conceived as an organic whole whose parts cannot be separated, so that each individual act contributes to the whole and itself originates in the central concept, then it will be perfectly clear and certain that the

supreme standpoint for the conduct of war, the point of view that determines its main lines of action, can only be that of policy. (Ibid., VIII/6: 607)

In Book I, Clausewitz had stated that action on the part of the defender is directed at changing the policy/political objectives of the attacker, and in Book VI that the attacker is more affected by policy/politics than the defender. In the earlier books of *On War,* the question of what constitutes politics/policy is not clearly defined—a reader could presume that Clausewitz's term means nothing more than imperatives prompted by rational consideration of national and international affairs. But in Book VIII he categorically rejects this formulation. Politics/policy is the product of deliberations by leading individuals, whose interests may or may not coincide with those of the state or nation, and whose actions therefore may or may not be rational in terms of national and international politics/policy. "Policy, of course, is nothing in itself," Clausewitz maintains,

> it is simply the trustee for all these interests against the outside world. That it can err, subserve the ambitions, private interests, and vanity of those in power, is neither here nor there. In no sense can the art of war ever be regarded as the preceptor of policy, and here we can only treat policy as representative of all interests of the community. (Ibid., VIII/ 6: 606–607)

Clausewitz offers further important clarification of his thinking when he insists that strategy is always subject to the rule of politics/policy, whether the latter is right or wrong:

> No major proposal required for war can be worked out in ignorance of political factors; and when people talk, as they often do, about harmful political influence on the management of war, they are not really saying what they mean. Their quarrel should be with the policy itself, not with its influence. If the policy is right—that is, successful— any intentional effect it has on the conduct of the war can only be to the good. If it has the opposite effect the policy itself is wrong. (Ibid., VIII/6: 608)

In the fourth chapter of Book VIII, Clausewitz addresses the concept of "center of gravity" with respect to guerrilla warfare—for Clausewitz a critically important instrument of defense—in a manner that clarifies and amplifies his previous discussion of both matters. In Book VI, he had defined the term as a state "where the mass is concentrated most densely," which in war is usually—but not always—represented by the concentrated military forces of each side. He noted that when this is the case, decision is most likely to be achieved by battle between the opposing centers of gravity—that is, the opposing concentrated armies. He observed, however, that the effectiveness of a blow struck by concentrated forces depends upon the object of such action also being concentrated. If this is not the case, "a blow may well be stronger than the resistance requires, and in that case it may strike nothing but air, and so be a waste of energy" (ibid., VI/27: 485–486). Clausewitz did not connect this observation to People's War explicitly, although the conditions of such a conflict match his point. But in Book VIII, Clausewitz maintains that in the case of "popular uprisings," the center of gravity is not an army, but "the personalities of the leaders and public opinion" (ibid., VIII/4: 596). Or, in other words, here Clausewitz makes it clear that the center of gravity of a defender waging guerrilla war is political rather than military, and as such insusceptible to destruction by concentrated military force alone. Such a characteristic, Clausewitz undoubtedly believed, is an important aspect of his contention that the defense is a stronger form of war than the offense.

Clausewitz's view that the defense is superior to the attack remains evident in the final chapter of Book VIII. "We regard the disadvantages that attach to the offensive," Clausewitz observes at one point, "as unavoidable evils." He later declares: "We maintain that the 1812 campaign failed because the Russian government kept its nerve and the people remained loyal and steadfast." The chapter ends with a description of a plan for a coalition war against a France that had resumed "that insolent behavior with which she has burdened Europe for a hundred and fifty years"—that is, a militarist France with expansionist ambitions (ibid., VIII/9: 627–628, 636). The fact that the plan called for an offensive war requires explanation.

In the summer and fall of 1830, the overthrow of governments in France and Belgium, as well as insurrections in parts of Germany, Italy, Switzerland, and Poland, appeared to presage another round of revolutionary military adventurism, which Clausewitz greatly feared. By the summer of 1830 he was convinced that war was inevitable, and in late 1830 he drafted proposals for both offensive and defensive action against France by means of a coalition consisting of Great Britain, Prussia, Austria, the Netherlands, and minor German powers. By early 1831, however, Clausewitz had concluded that general circumstances favored a defensive strategy.[11] The argument for an offensive in Book VIII, which was perhaps written before the end of 1830,[12] appears to reflect his earlier position—that is, one that had developed during a period when he was tempted by the possibilities of the attack. Even at this time, Clausewitz may have harbored doubts about the viability of an offensive strategy by a coalition that—minus Russia—was considerably weaker than the one that had barely prevailed against France in 1814. And by 1830, France was in an even better position than it had been in 1814, a time when its strength had been reduced by years of military exertion and disaster.

The question of whether Prussia and its allies should initiate hostilities against France in 1830 and 1831 posed an enormous challenge for Clausewitz, his capacity for clear thinking and command of the appropriate theoretical instruments notwithstanding. That even he, for a time, could express support for a course of action that violated his core convictions is a measure of the difficulties that decision-makers face. For those inexperienced in war, Clausewitz observed, choosing among strategic options may appear easy enough. Reality was different. "Everything in war is very simple," he warned, "but the simplest thing is difficult" (*On War*, I/7: 119).[13]

Conclusions

ARL VON CLAUSEWITZ POSSESSED the attributes of a great scholar, but he was not an academic. His mind was penetrating, rigorous, contemplative, and highly literate. Yet he was also a soldier who fought for his country's survival in wars of unprecedented scale and ferocity. Thus, although the Prussian author was by talent and temperament an intellectual, for a time his life was one of extraordinary action—on the battlefield, from headquarters, and in strategic deliberations at court. What he had to say to the world was not drawn principally from books or from discourse with learned colleagues, but from his own observations and reflections on personal combat, supreme command, and high politics. Clausewitz's ideas were politically and militarily too unconventional to find favor with those in power following the wars against Napoleon. Unable to influence directly the course of events or to play a constructive role in Prussian military education, he turned to writing a book that addressed posterity. Clausewitz undoubtedly believed that his work had things to say about war that were of lasting general significance. But his driving concern was to promote views on strategy and generalship that he believed were of immediate vital importance to his country's security.

Military necessity was the mother of Clausewitzian invention. In 1806, the high command of Prussia's army was both overaged and unseasoned, while that of France combined youth and experience. The wide disparity in the competence of leadership at the top contributed in important ways to Prussia's catastrophic defeat. The battlefield debacle and the political demoralization that followed caused the Prussian crown to pursue a policy of cooperation with France, which not only was an affront to Clausewitz's patriotism but courted national destruction. Prussia's fortunes were restored in the wake of French overextension and defeat by a coalition. Being on the winning side in 1815, however, did not change the fact that Prussia remained much inferior to France in terms of size, population, and military power. Prussia's weakness and proximity to its arch-enemy made it a likely target in the event of a resurgence of French political radicalism and military aggression, which seemed imminent in the late 1820s. In *On War*, Clausewitz thus addressed the problems of supreme command and strategic perspective to prepare the leadership of his country for a repetition of the French onslaught of 1806. He proposed a novel form of studying the conduct of war as a means of improving the capability of a future commander-in-chief. He also explained how and why effective national defense is possible even after military disaster, intending to provide a tonic to Prussian strategic equanimity in the event of such an occurrence.

Clausewitz's pedagogical creativity was prompted by his dissatisfaction with the prevalent practice of using principles derived from history as the basis of officer education. He was convinced that the conduct of war was too complex and unpredictable to be based—as Jomini would have it—upon military maxims. Clausewitz also knew, from his personal observation of senior leaders, that executive capacity in war could be improved greatly by experience. But the absence of major conflict for a decade or more—as had been the case in 1806 and after 1815—ruled out the possibility of a Prussian army being led by an officer who had exercised supreme command in a major war and who also possessed the vitality of youth. Clausewitz thus conceived of an educational procedure that reproduced the emotional as well as the intellectual difficulties of supreme command, which could serve as a substitute for actual experience. This is the imagined replication of command decision in the past—that is, historical reenactment. The advantage of historical reenactment over role-playing a hypothetical case is that the former deals with what is believed to be reality, which provides a stronger basis for emotional involvement. In addition, historical reenactment does not require one to make up the governing conditions that would be known about a real event in the past, which Clausewitz would have regarded as a dangerously arbitrary proceeding.

A major drawback of reenactment, however, is that the historical record is incomplete. That is to say, verifiable information about a host of factors that affected the decision-making of the commander-in-chief in particular circumstances—many of which constituted what Clausewitz called "general friction"—does not exist. Clausewitz thus argues that historical facts must be augmented by surmise about what could not be known. Surmise is not to be a matter of pure speculation, but based upon propositions about the probable nature of the issues in question derived from personal involvement in—and observation of—the exercise of supreme command. Clausewitz regards these propositions collectively as a body of theory.

Clausewitz intends such theory to supplement known historical data in six ways. First, theory directs attention to the factors that promote self-doubt in the commander, which include danger, complexity, contingency, and the unreliability of information about what is going on and to the emotional resources needed to counter them. Second, it supports conjecture about the factors that

Conclusions

inform the commander's judgment, which encompasses his knowledge of policy and politics, assessments of people and issues, and comprehension of the quality of the forces commanded. Third, theory provides the basis for the consideration of the multitude of operational facts and motives for action of many individuals that are either never known, or if known, never recorded, or that may have been intentionally obscured. Fourth, it makes certain that proper account is taken of the nature of the relationship between cause and effect in war, which is affected by the play of unintended consequences and complexity. Fifth, theory mandates consideration of alternative courses of action as an essential part of the process of replicating command dilemmas. And sixth, it allows knowledge of outcomes (that is, the success or failure of the operation) to influence surmise about the roles of the unknowable variables just described and their complex interactions when evaluating the rightness or wrongness of decision-making.

In Clausewitzian reenactment, historical authenticity is less important than intellectual and emotional verisimilitude. This is because the aim of reenactment is not imitation of the behavior of the historical actor, but replication of conditions of decision-making that pose comparable, if not the actual, intellectual and moral challenges of the historical case. The product of such a substitute for actual experience is supposed to be improved intuition, which Clausewitz regarded as, among other things, the main antidote to the negative effects of general friction. Decision-making by the reenactor can follow that of the historical commander or take a different course. But in either instance, the reenactor must assume responsibility for his decisions in order to recreate a kind of intellectual and moral experience upon which he can reflect and learn. For Clausewitz, the purpose of theory with respect to history is not representation of conclusive arguments about war, which he believes to be of little practical significance, but augmentation of the ability of history to recreate a past psychological reality. Theory of this kind is a facilitator of an individualized learning process, not a prescriptive instrument, and thus valid for all kinds of war, ranging from the least to the most intense in violence.

Clausewitz's second major theoretical construction is his contention that the defense is the stronger form of war. For Clausewitz, successful resistance to invasion is possible even when the attacker is much stronger than the defender. He supports this general contention with two main subordinate propositions— that the conduct of war is shaped by political/policy considerations at all times, and that politics/policy affects the attacker more than the defender. Although

Conclusions

Clausewitz formulated these ideas with the strategic relationship of France as an attacker and Prussia as the defender in mind, he believes them to be valid for any situation. His positions do not prescribe action; rather, they describe certain governing dynamics—henceforward referred to as what we shall call "the nature of things"—that are supposed to be taken into account when reenacting command decision. Indeed, Clausewitz regards the greater strength of the defense over the offense as the main reason for the suspension of action in war. It was, for this reason, a major source of strategic dilemma.

Clausewitz maintains that his arguments work in combination. In war, the political/policy motive of the attacker—to compel the defender to act against its interests—is opposed to the political/policy motive of the defender—to discourage the attacker. In addition, the political/policy motives of both sides are conditioned by internal political considerations, which are a matter of the degree of agreement or disagreement within governing circles or between governors and governed. As a general rule, the energy required to sustain an offensive is greater than that required to maintain a defense. This is especially true when topography favors defensive fighting or when expansive territory allows the defender to retreat to buy time. All else being equal, an attacker will likely reach critical thresholds of internal political difficulty over the escalating costs and risks of war before a defender does. And such is the extent of the disproportionate costs of the attack compared to the defense that this is so even when the attacker is considerably stronger than the defender. In either case, the ultimate effect of political crisis is to reduce the attacker's aspirations and thereby bring hostilities to a close.

In a war in which the objective of the attacker is the destruction of the defender's sovereignty, the difficulties for the attacker are increased by the inherently greater strength of the defender's political/policy motive. This is because the moral stakes for the defender are about existence, which is essential, whereas the attacker is concerned simply with gain, which is discretionary. Moreover, the resources available to the defense for military action can overmatch those of the attacker if the defender government's will to resist enjoys broad internal political support. Under these circumstances, the regular forces of the defender can be augmented by the armed action of an aroused citizenry—that is, guerrilla war—while the attacker cannot count on involvement from its own civilian population. A defender that has demonstrated a determination to resist even a greatly superior attacker can also count on the assistance of other powers, which are likely to recognize that their own independence is threatened by the offensive success of a state with aggrandizing or even hegemonic intentions. In short, effective defense against attack is not just about military action, but about the interplay between military performance and a variety of internal and external political dynamics. This is probably what Clausewitz had foremost in his mind when he stated that "war is simply a continuation of political intercourse, with the addition of other means" (*On War*, VIII/6: 605).[1]

Clausewitz's views on defense challenge the universal applicability of Jomini's cardinal principle of war: concentration of force. Clausewitz recognizes that concentration of force is desirable, if not essential, to winning battles,[2] but he does not believe that victory in battles determines the outcome of all wars. In cases of conflicts in which the sovereignty of the defender is at stake, Clausewitz maintains that even when the concentration of greatly superior forces results in the destruction of the defender army, this is not enough to end hostilities if the defender possesses the will to continue fighting by all available means, including guerrilla war. Concentration of force on the part of the defender is required to achieve decisive victory through counterattack, a course Clausewitz favors whenever circumstances permit. He insists, however, that a defender that is too weak to launch an offensive can still obtain favorable terms by discouraging the attacker through protraction of hostilities. Thus, given defender will to resist at all costs, decisive battle is unobtainable for an attacker, and strategic victory highly unlikely or even impossible. This is not to say that the offense will inevitably fail, but rather that the balance of military force is not the critical strategic variable. Instead, what matters is the relative strength of the attacker and the defender's determination.

Clausewitz does not believe that any theoretical formulation, including his own theoretical statements on the relative strengths of the defense and attack, can prescribe the actual conduct of war. But this does not rule out the use of theoretical propositions to set the terms of thinking about a strategic problem. Theory accomplishes this by identifying the nature of things in war. By so doing, it pushes deliberation in directions that it might otherwise not have gone, raising questions rather than providing answers. The purpose of such a process is to prevent bad intellectual habits—such as maintaining belief in the decisive strategic significance of concentration of force—from determining strategic courses of action. In the specific case of attack and defense, the point of a potential attacker contemplating the superiority of the defense over the offense is not to learn to reject offensive action, but to be able to consider the strategic implications of fighting a defender that possesses the will to protract the war. Conversely, from the defender's point of view such an exercise provides an opportunity to consider protraction of hostilities as a practicable alternative to surrender in cases of catastrophic military defeat and occupation. Or, to put it in more general terms, Clausewitz's dictum is supposed to counteract any predilection on the part of either the attacker or the defender to believe that a very great military success at the outset of hostilities is ipso facto tantamount to a political decision.

Between 1827 and 1831, Clausewitz had four years to turn an incomplete draft manuscript, which he characterized as "a shapeless mass of ideas" that would be "liable to endless misinterpretation," into a nearly finished book (ibid., 70). Given the fact of available time and the character of the known text, there can be little doubt that at Clausewitz's death, *On War* was complete with respect to its general form and major arguments. Had Clausewitz lived to finish his work, we can assume he would have modified details and refined the way his ideas were expressed, but there is no reason to believe that the overall structure or the main substance of his reasoning would have changed. *On War* cannot, therefore, be dismissed as unintelligible because it is unfinished.

Accurate comprehension of Clausewitz's treatise has nevertheless escaped all authoritative commentators. There are three reasons for this. First, Clausewitz's main arguments are difficult and complex. Second, the exposition of the arguments is not always clear, making the book prone to misinterpretation. And third, inaccurate preconceptions have interfered with the proper negotiation of the text. The complexity of Clausewitzian argument is a function of two conditions: the presentation of many propositions of greater and lesser importance, and their intricate interconnection. Clausewitz's main arguments can be divided into two categories—statements about what can be called the "essential characteristics of sound general theory," and statements that address practical problems. Arguments in the first category define Clausewitz's criteria for a valid approach to the study of the conduct of war. Arguments in the second category address officer education and the relative strengths of strategic defense and offense.

Clausewitz has six arguments related to the essential characteristics of sound general theory. First, a general theory of war must be valid for all forms of war, from the least to the most intense in violence. Second, a general theory of war must deal with all major and minor factors that affect action in armed conflict in terms commensurate to their effects, including both the moral and

Conclusions

material aspects of war, and must also deal with the large and all-pervasive influence of political/policy factors on decision-making at the strategic level. Third, a general theory of war must account for suspensions of action that characterize most wars. Fourth, a general theory must recognize the possibility, if not probability, of large-scale battle—that is, no sound general theory of war can be formulated in which at least the potential for a large-scale engagement does not exist. Fifth, a general theory of war must address the *study* of the conduct of war, not the *actual* conduct of war; that is, the purpose of theory is to improve the process of examining history, a procedure that Clausewitz believes could be made scientific, not the prescription of action, which he holds can never be. And sixth, a general theory of war must recognize that armed conflict does not exist within the realm of the arts and sciences; instead, it is an aspect of man's social existence. The effective conduct of war is therefore less a matter of mastering a set of specialized techniques and more one of a combining self-knowledge, knowledge of others, and a sense of the nature of things in war, with decision-making ability in complex and stressful circumstances.

With respect to practical problems, Clausewitz puts forward fundamental propositions about general matters, presents revolutionary ideas, and adds supportive reasoning. Insofar as fundamental propositions about general matters are concerned, he states that decision-making by the commander-in-chief is the single most important factor in the execution of military strategy, and that the proper education of officers with respect to the problem of decision-making at the strategic level is therefore crucial to the state. The two revolutionary ideas are that officer education should be based upon the historical reenactment of command decision, using a novel method of combining history and theory, and that at the strategic level, the defense is the stronger form of war. The three supportive arguments are that politics/policy shapes the ends of strategy during the war as well as before its outbreak, that policy/politics affects the attacker more than the defender, and that guerrilla war is an essential defensive instrument of last resort for any state threatened with destruction by a more powerful opponent. The fundamental propositions about general matters constitute the points of departure for the revolutionary arguments, and the supportive arguments are essential components of both revolutionary arguments.

The difficulty of understanding Clausewitzian argument has a great deal to do with the fact that it takes the form of a theory of practice rather than a theory of a phenomenon. Presuming that it is the latter, as is almost universally the case, transforms instruction in how to learn into instruction in how to behave.

In making this error, readers adopt expectations of integrated descriptive order and prescription that are alien to the intent of the author. And even recognition that *On War* is a theory of practice is not in and of itself sufficient to ensure proper comprehension of Clausewitzian thought. Raymond Aron and W. B. Gallie nominally acknowledged that *On War* offered a theory of practice, but they did not explore or develop this view.[3] And no major critic has recognized that the crucial factor in Clausewitz's conception of practice is intuition.

Clausewitz is convinced that intuition is the primary agent of decision in the face of command dilemma in war. Deliberate reasoning, in his view, is insufficient in the face of the complexity and incomplete information existing under conditions of contingency and danger. The functioning of intuition in war combines two kinds of dynamics: subrational reasoning and emotional equanimity. Neither faculty can be understood in terms of rules or principles, which are useless with respect to their development. Clausewitz offers two solutions to this problem: a method of grappling with the intractable directly through reenactment of individual historical cases, and a statement about the superiority of the defense over the offense that violates certain conventional attitudes toward war in a way that can be regarded as counterintuitive. The former encourages the development of sound intuition and the capacity to deploy it, while the latter serves as an antidote to the intellectual elements that form the basis of what could be called unsound intuition. Undoing defective habits of thought requires the ability to engage in serious critical self-examination, a task that is never easy. Accepting the necessity of reenactment demands some appreciation of what constitutes a sound scientific approach to the study of war, the limitations of language as a medium of direct communication, and certain sophisticated concepts about the nature of historical inquiry.

Clausewitz believed that a scientific approach is about the observation of the particular characteristics of phenomena, not the operation of general laws or principles. The use of general laws or principles is allowed, but only to facilitate a more comprehensive and accurate observation of the particular than might otherwise have been the case. He considers the use of general laws or principles to explain outcomes an illegitimate use of theory, which is to say unscientific. Such explanations almost always—without justification—reduce complex causation to something simpler. The distinction between sound and unsound use of general laws and principles in inquiry that claims to have a scientific character is also a major concern of Charles Sanders Peirce, whose views on the subject are indicative of its intricacies.

Conclusions

Clausewitz recognizes that language alone is incapable of describing the conduct of war in a manner that can produce rational and emotional understanding equivalent to the effects of actual experience. This is because he, like Ludwig Wittgenstein a century later, believes that words can convey little more than a crude approximation of any complex and difficult reality, especially when a large part of experiencing that reality involves the play of emotion. That said, Clausewitz had no better medium of communication for his purposes than language. He was thus compelled to devise a means of using words to induce understandings that could not be explicated directly. His solution, as has been explained, was to construct something that resembles experience — to the point of being nearly equivalent to it — through historical reenactment based on historical fact and theoretically informed surmise. Coming to terms with the linguistic philosophical reasoning that drove Clausewitz to such an expedient, even in the light of later philosophical writing making similar points about language, constitutes a large intellectual challenge.

Clausewitz's concept of mental historical reenactment to a remarkable degree anticipates the ideas of R. G. Collingwood, one of the twentieth-century's leading students of the philosophy of history. Historical reenactment is one of Collingwood's signature ideas, and his explanation of it specifically addresses the subject of writing strategic and military history. Collingwood's description of historical reenactment and his justification for it have generated a considerable critical debate, which may serve as an indicator of the difficulty and complexity of what might at first glance appear to be a relatively simple notion. Clausewitz's form of historical reenactment, moreover, contains an element missing from Collingwood's proposal, namely, the necessity of using theory-based surmise to make up for gaps in the historical record. This major proposition further complicates an already philosophically fraught enterprise.

The complexity and difficulty of Clausewitz's arguments are compounded by shortcomings in his exposition, particularly in his discussions about historical reenactment, absolute and real war, general friction, and the relationship between war and politics. Misleading nomenclature obscures the presentation of historical reenactment. Clausewitz calls the complex process of reenactment and reflection on reenactment "critical analysis," a term that focuses attention on the evaluative component rather than on producing a narrative to form the basis of a procedure. Thus prompted, a reader can easily suppose that Clausewitzian theory means nothing more than an enlargement of the categories that are to be taken into consideration when judging the rightness or wrongness of command decision to include nonmaterial and otherwise unquantifiable factors that were previously ignored. Or a reader might believe that Clausewitzian theory is simply a corrective to the supposedly "uncritical analysis" characteristic of conventional prescriptive approaches to strategic and tactical case study, instead of seeing it as providing components of a story line whose pedagogical function is nonprescriptive.

Clausewitz's concepts of "absolute war" and "real war" have provoked a great deal of confusion. The fundamental problem is his initial presentation of absolute war as no more than an abstraction. Clausewitz subsequently qualifies this formulation by recognizing the existence of real conflict that approximates absolute war, namely fighting driven by both the propensities to maximize violence and the pursuit of political objectives. But Clausewitz's definitional shift is not explicit, and thus requires attentive reading to detect. The effect of lack of clarity on this matter is compounded by Clausewitz's contention that during hostilities, the categories of what has been called "(less than absolute) real war" and "(real) absolute war" are in effect conflated. Before the achievement of a formal peace, Clausewitz explains, there is always the possibility that the former form of war will be transformed by political circumstances into the latter form through the escalation of violence. For this reason (less than absolute) real war and (real) absolute war are not separate and opposed categories, but different states of the same phenomenon. And it is this condition, and not the characterization of war as either one sort or the other, that Clausewitz believes should govern the direction of action by the commander-in-chief (and thus the reader as reenactor).

Clausewitz conceded that describing friction in its multiple forms would "take volumes to cover," and that it was "a force that theory can never quite define" (*On War*, I/7: 120). The concept is nonetheless described clearly enough in outline in Book I. He unquestionably believes that friction is of major theoretical significance, stating that "we shall frequently revert to this subject" (ibid., I/7: 121). In subsequent chapters, friction and general friction (of which friction is a subset) are often addressed as promised. Neither term, however, is used in the course of explaining historical reenactment in Book II, even though both concepts help to make theory-based surmise necessary, because of the lack of verifiable evidence of their effects in particular cases. The connection of friction and general friction to Clausewitz's arguments on historical reenactment is thus obscured. And as a consequence, readers are more likely to view friction and general friction as taxonomic categories in a theory of a phenomenon than to recognize them as critical active agents in a theory of practice.

Conclusions

To understand Clausewitz's statement about the defense as the stronger form of war the reader must make a connection between the main argument and its subordinate propositions: that "war is an extension of politics/policy [at all times]"; that "politics/policy affects the attacker more than the defender"; and that guerrilla war should be accepted as a decisive defensive instrument of last resort. Clausewitz's discursive explication of the main and supporting arguments, however—which is divided among Books I, VI, VII, and VIII—is extremely difficult to follow. This may be the case because the political/policy aspects of his views on the superiority of the defense over the offense were a late addition to the manuscript, as indicated by the note of 10 July 1827. Clausewitz—given the time for a final revision—might have produced a better-integrated presentation of his interlocking arguments and improved the transparency of the explanation with forthright introductory and summary statements. Or Clausewitz's failure to make an explicit connection between guerrilla and (real) absolute war—which is implicit and crucial to a full understanding of his concept of the relationship between war and politics—may have been due to his desire to avoid provocation of conservative and, in particular, royal displeasure. Whatever the reason for the complicated structure of his presentation, it makes the relationships between the separate arguments hard to discern.

Finally, access to the meaning of *On War* has been obstructed by erroneous preconceptions in much of the critical writing on Clausewitz in the nineteenth and twentieth centuries. In addition to setting out faulty and incomplete analyses of Clausewitzian argument, the character and causes of which have been covered in our previous discussion, this literature produced debilitating prejudice by prompting the acceptance of mistaken attitudes about theory and the meaning of certain key terms.

As explained by Alan Beyerchen, consideration of military theory has been almost universally corrupted by the assumption that war is a near-linear phenomenon that can be dealt with adequately by theory that provides linear approximations that are known as principles of war. Jomini established this approach as standard, and his intellectual heirs ensured its perpetuation in the curricula of war colleges and in the discourse of scholars and pundits. Readers conditioned by this perspective are almost bound to see Clausewitz's criticisms of Jomini as nothing more than a corrective to the *rigidity* of Jominian determinism, rather than as a rejection of determinist theory as such, and to miss altogether the nature of Clausewitz's solution to the problem of the nonlinear character of war, namely, a nonprescriptive theory of practice that promotes the development of sound intuition and discourages unsound intuition.

Yet another kind of preconception pertains to the misunderstanding of certain terms used in *On War* because of the widespread practice of viewing the text in light of later strategic concerns. Clausewitz sees guerrilla war as a form of (real) absolute war. For Clausewitz (real) absolute and (less than absolute) real war refer to the character of armed conflict in terms of the magnitude of violence, whereas his concepts of "total defeat of the enemy" and "war of limited aim" deal with the magnitude of the strategic objective (ibid., VIII/7: 611). The two sets of concepts, while related, are not interchangeable—that is to say, a "war of limited aim" might not always be a (less than absolute) real war, and a war whose goal is the "total defeat of the enemy" is not always the equivalent of (absolute) real war. A defender that resorts to guerrilla war with the objective of causing political discouragement for the attacker, rather than the destruction of the attacking army, can be said to be fighting a limited (real) absolute war.

In the twentieth century, however, the two sets of Clausewitzian terminology have been conflated, with (real) absolute war being regarded as the same as what is called unlimited war, and (less than absolute) real war the same as what is called limited war. In the parlance of the 1960s and after, unlimited war refers to a full-blown exchange of atomic weapons with an effect tantamount to the total destruction of the combatants, if not the world; limited war refers to restrained use of nuclear weapons or any use of conventional weapons; and guerrilla war to a low-intensity variant of the latter. The characterization of guerrilla war in this way, when applied to a reading of the text of *On War*, confuses what Clausewitz means by absolute war as a form of real war and makes it difficult to see why guerrilla war as a manifestation of (real) absolute war is a critically important component of his theoretical outlook.

Certain assessments of the Vietnam War heavily influenced the interpretation of Clausewitz's statement that war is an extension of politics/policy. Critics of American conduct of the war attributed its lack of success in large part to the failure of the national leadership to formulate clear and attainable political objectives as a basis for military action, citing Clausewitz's dictum as their justification for this judgment. The reciprocal effect of such usage established the practice of interpreting the phrase as meaning no more than that the decision to resort to military action and military action itself should be governed by political/policy considerations. As a consequence, the American understanding of the phrase excluded the several other implications of Clausewitz's statement: that

Conclusions

political/policy considerations can alter over the course of hostilities; that politics/policy includes the impact of internal politics as well as the external goals of national action; that because of internal politics the attacker is more adversely affected by politics/policy than the defender; that the objective of the defender—in the absence of the ability to defeat an attacker militarily—is to exploit this propensity to its advantage; and that the ultimate instrument of defensive resistance is guerrilla war. As a result, Clausewitz's famous phrase has been divorced from the associated propositions that give it the more complex and nuanced meaning intended by the author. Writers on Clausewitz have thus failed to recognize this statement as a critical component of one of his two revolutionary arguments, again, that the defense is the stronger form of war.

The cumulative effect of the errors induced by the complexity, difficulty, and imperfect expression of Clausewitz's arguments and the erroneous preconceptions of scholars has been to foster a methodological consensus that explicitly or implicitly characterizes *On War* as an incomplete collection of fragments. Study has been largely focused on the elucidation of the meaning of particular fragments rather than explication of the meaning of the whole. The task of coming to terms with the meaning of the whole, moreover, is either considered to be futile or put off to an indefinite future pending further study of fragments. In this book I have tried to identify Clausewitz's specific theoretical objectives and the novel means by which he achieved them in an attempt to show that *On War* is much more than a collection of fragments; indeed, it is a bold and innovative approach to the study of war that weaves together themes of critical importance in a way that anticipates some of the greatest philosophical thought of the nineteenth and twentieth centuries. Seeing it as a coherent whole is a necessary prerequisite to dealing with the "fragments"—that is, the details that Clausewitz incorporates are further evidence and illustration of his main concepts. I have therefore used broad strokes in order to show the big picture. My goal was not to produce a conclusive interpretation of the entire substance of *On War,* but to establish a viable point of departure for intelligent reading of this great work by ordinary civilians, military professionals, and serious scholars.

Clausewitz maintained that the conduct of war could not be directed by knowledge of principles, insisting that the critical strategic factor in war is a matter of the balance of political determination, not military force. To think otherwise, he was convinced, is to promote faulty observation of history, to corrupt reflection, and to encourage self-delusion. Clausewitz believed that his own experience and study of history had enabled him to comprehend the dynamics of supreme command, which, he held, was the crucial variable with respect to the proper use of active armed forces in international politics. In *On War*, Clausewitz explains how a person who lacks actual experience of directing an army under difficult circumstances can acquire such an understanding, gaining a sensibility rather than a form of knowledge. To assess the continued viability of his great book as a guide to the study of war, therefore, we must evaluate two things: the continued relevance of his experience and historical findings, and the soundness of his conception of what and how to learn. I will thus conclude with a review of the validity of Clausewitz's experience and studies and the testimony of great thinkers who shared Clausewitz's attitudes about both the nature of man as a decision-maker and serious education.

In the present, as in Clausewitz's day, history is used in military education primarily for two purposes: to teach lessons or to illustrate the operation of supposedly immutable principles of war. In both cases, theory is used to explain the totality of causation, and the main concern is with what constitutes good as opposed to bad conduct. Clausewitz believed that such an approach to case study was essentially ahistorical—and thus unscientific—because actual decision-making at the strategic level was not a matter of making a choice between what was thought to be correct as opposed to incorrect, but of overcoming dilemmas generated by difficult, complicated, and changing circumstances without adequate information. The proper study of the past, Clausewitz concluded, required consideration of why decisions were hard, rather than whether they had been right or wrong. His proposed alternative to conventional practice was historical reenactment, a procedure that combined historical fact with theory to construct narratives that took into account important but unknowable things. The failure in the past to comprehend Clausewitz's analysis of the strategic decision-making process and his method of coming to an understanding of its nature, and the fact that his line of reasoning and conclusions were not invented by anyone else, means that *On War* has retained its power to offer a productive approach to the study of armed conflict.

In *On War*, theory that is needed to imagine unknowable factors essential to productive historical reenactment is derived from observation of command decision in armed conflicts of Clausewitz's time and place. This aspect of his

Conclusions

methodology is thus characterized by historical and geographic particularity. Although arguably valid for case study situated in the late eighteenth and early nineteenth century in Europe, it must be in important respects inappropriate or incomplete for other periods and locales. Applying the concept of historical re-enactment to twentieth and twenty-first century case studies will therefore require the creation of an updated body of theory to extend, correct, or otherwise modify the theoretical materials on strategy, tactics, and general friction in Books III, IV, and V. In addition to the recent experience of major practitioners or observers of command decision, the generation of intelligent surmise about unknowable factors of critical importance will necessitate consideration of the findings of political science, sociology, anthropology, and psychology, with the assistance of philosophy and cognitive science as well. The creation of an educationally productive historical narrative in the fullest Clausewitzian sense, therefore, will be a multidisciplinary enterprise drawing upon the resources of the biological and social sciences as well as the humanities.

Clausewitz's contention that the defense is the stronger form of war, and its supporting propositions on the relationship between war and politics and the importance of guerrilla war, have been validated by the military history of the nineteenth and twentieth centuries. This question requires careful consideration, which must begin with a review of the general dynamics of the Wars of the French Revolution and Empire. From 1805 to 1812, a combination of superior manpower and fighting technique, and extraordinary leadership at the top, enabled France to win wars that extended its territory and influence. But tactical and operational virtuosity were ultimately not enough to produce strategic victory in the face of the resistance of governments and peoples who had come to believe that French expansionism threatened their nationhood. In Spain, a combination of regular troops, the expeditionary force of an ally, and armed civilians acting over several years inflicted prohibitively high losses on the occupying French army. In Russia, territorial overextension, compounded by the effects of winter and popular resistance, resulted in the near complete destruction of the French invading forces within one campaign season. In Germany, a national uprising against French occupation replenished the ranks of the once-shattered and subsequently much-reduced Prussian army, which then returned to the field as a formidable fighting force. These defensive successes set the stage for a coalition counteroffensive that brought about the collapse of French power.

But events that followed Clausewitz's death appeared to contradict his argument that the defense is the stronger form of war. In 1866 and 1870, Prussian military offensives against nominally stronger great-power rivals achieved decisive victories quickly, and the consequence was the unification of Germany under Prussian leadership. This made Prussia, the leading element of the German empire, the preeminent military power on the continent. This change had two effects on the perception of the practical applicability of Clausewitz's views. In the first place, the demotion of France to the position of a militarily inferior power eliminated the prospect of French invasion and occupation as Prussia's primary national security problem, which relieved Clausewitzian theory of the burden of addressing a replay of the events of 1806. And second, Prussian victories in the wars of German unification seemed to demonstrate that a properly executed strategic military offensive could be politically productive. Ironically, the very dissipation of Clausewitz's greatest fear called into question the soundness of a theoretical proposition that is of central importance in *On War*.

In fact, the strategic circumstances of the nineteenth and twentieth centuries support rather than invalidate Clausewitz's position on the relative merits of the defense and the attack. His views had been based upon the assumption that contending armies are essentially similar in size, form, weaponry, and command apparatus, and formulated primarily to address cases in which the objective of the attacking power is the destruction of the sovereignty of the defender. These conditions were to a significant degree not operative in mid-nineteenth-century Europe. During this period, Prussia enjoyed substantial superiorities over its opponents in military organization, armament, and the structure of leadership and pursued political goals that did not threaten the national existence of its enemies. And although spectacular political success crowned Prussian military offensive achievement in wars with Austria and France, French protraction of the war through resort to guerrilla warfare and fighting by untrained levies caused serious disagreements within the German leadership that very nearly had large consequences.[4]

In the first half of the twentieth century, Germany attempted twice to achieve hegemonic political objectives through offensive strategic action. But although the German army was qualitatively superior to those of its opponents, it proved incapable of producing rapid decisive victory. Its early operational successes in both world wars were thus contained, with total defeat at the hands of a superior coalition following after extended hostilities. Prussia's transformation

Conclusions

into Germany in the nineteenth century had freed it from the condition of dire insecurity, but not from the dynamics of strategic reality as described by the Clausewitzian imperative that the defense is a stronger form of war than the offense.

In the nineteenth century, great wars of national survival did not occur in Europe, so naturally, guerrilla warfare did not emerge as a major form of fighting in the Clausewitzian sense. And in 1918 and 1945, internal division and demoralization in Germany precluded resort to guerrilla warfare as a response to invasion by its counterattacking enemies, as had also been true of France in 1814 and 1815. After World War II, the outcomes of conflicts in Algeria, Vietnam, and Afghanistan, however, demonstrated that insurgent forces with the will to sustain heavy losses were capable of discouraging much stronger opponents. In these conflicts, the dynamics of guerrilla warfare were as Clausewitz had stated: An attacker—defined as a state attempting to impose its will on a national population other than its own—is thwarted when the protraction of hostilities precipitates internal and external political difficulties for the attacker that, in combination with military losses, are sufficient to cause a modification of political objectives, namely, the abandonment of the offensive effort. In sum, Clausewitz's assessment of the relative strengths of the defense and the offense, when considered in light of history, appears to be sound.

In *On War*, Clausewitz maintains that the study of armed conflict should be focused on the human faculties that constitute the agent of effective command decision. In certain critical respects, the conditions of such decision-making resemble those characteristic of the major vicissitudes of ordinary life. In both cases, responsible individuals are confronted by hard and complex problems that must be solved in the face of uncertainty and peril. This is to say that with respect to intellectual and moral demands, war is like peace, only much more so. Major arguments in *On War* can thus be related to serious writing about life in general when it is concerned with the nature of human dilemma, the role of self-transformation as a means of developing the capacity to make decisions in spite of it, and the relevance of historical learning to self-transformation. Three examples of such work are presented here in order to broaden the terms of appreciation of Clausewitzian thought.

In *On War*, Clausewitz considered the strengths and limitations of individual man as an executor of decision under difficult circumstances. Alexander Pope (1688–1744) contemplated this matter in a well-known work that could have been known to Clausewitz.[5] Nearly a century before the publication of *On War*, Pope had written,

Know then thyself, presume not God to scan
The proper study of mankind is Man.
Placed on this isthmus of a middle state,
A being darkly wise and rudely great:
With too much knowledge for the sceptic side,
With too much weakness for the stoic's pride,
He hangs between; in doubt to act, or rest;
In doubt to deem himself a god, or beast;
In doubt his mind or body to prefer;
Born but to die, and reasoning but to err;
Alike in ignorance, his reason such,
Whether he thinks too little, or too much;
Chaos of thought and passion, all confused;
Still by himself abused, or disabused;
Created half to rise, and half to fall;
Great lord of all things, yet a prey to all;
Sole judge of truth, in endless error hurled;
The glory, jest, and riddle of the world![6]

For Clausewitz, the improvement of a man's capacity to direct action requires the removal of certain impediments to what can be called personal enlightenment. Clausewitz may have been referring to the importance of this very concept when he stated in 1818 that his first attempt to develop a general theory on war had been informed by the approach of Montesquieu (1689–1755) in *The Spirit of Laws*.[7] Clausewitz did not explain what his exact views on this matter were, but the following quotation from the introduction to the French philosopher's great work is suggestive. "I would consider myself the happiest of mortals," writes Montesquieu, "if I could make it so that men were able to cure themselves of their prejudices. Here I call prejudices not what makes one unaware of certain things but what makes one unaware of oneself."[8]

For Clausewitz, in the absence of experience, the most effective means of dismantling the prejudice underpinning self-delusion is the intelligent study of the past. Such activity, he believes, is worthy in spite of its inherent limitations and indirect applicability to practical affairs and the fact that history is largely a record of human folly. Herman Hesse (1877–1962), the German novelist, seems to have had a similar outlook, which he expressed in *The Glass Bead Game*, which was published in the midst of World War II:

Conclusions

I have no quarrel with the student of history who brings to his work a touchingly childish, innocent faith in the power of our minds and our methods to order reality; but first and foremost he must respect the incomprehensible truth, reality, and uniqueness of events. Studying history, my friend, is no joke and no irresponsible game. To study history one must know in advance that one is attempting something fundamentally impossible, yet necessary and highly important. To study history means submitting to chaos and nevertheless retaining faith in order and meaning. It is a very serious task, . . . and possibly a tragic one.[9]

Appendix One
A Pictorial Representation of Critical Analysis

In Clausewitz's day and ours, analysis of military events in the past frequently involves the derivation of theory from the study of history. This approach consists of reflection upon a historical narrative constructed from verifiable historical fact (VHF), which generates a set of general propositions that are supposed to be valid for most but not all cases. The foregoing can be expressed as follows:

Verifiable Historical Fact (VHF) + Reflection on Verifiable Fact (RVHF) = Theory Qualified by Exception (TQE)

In this procedure, history and theory are separate, with the former constituting the source of the latter. This relationship can be illustrated as follows:

Clausewitz finds such an approach fundamentally unsound because he believes that verifiable historical fact never provides enough information for the theorist to reconstruct accurately the dynamics of past cases of strategic command decision. Reflection upon an unsatisfactory representation of past reality, in turn, results in theory that offers only an incomplete or otherwise flawed explanation, and for this reason such a procedure is of little practical value. Clausewitz considers activity of this kind to be "uncritical" analysis.

In Chapter 5 of Book II of *On War*, Clausewitz states that critical analysis of the dynamics of command-decision in the past consists of three things. The first two—historical research that clarifies matters of fact, and surmise about matters that must have affected command decision—are based on theory derived from personal observation of the exercise of supreme command in war. The product of both of these activities in combination provides the material for a mental reenactment of decision-making by a commander-in-chief that is meant to approximate, if not replicate, actual experience—that is, something equivalent to a past reality. The third is reflection on this synthetic experience,

Appendix One

the product of which can be described as an improved ability to make strategic decisions under difficult conditions.

The foregoing can be expressed as follows:

Verifiable Historical Fact (VHF) + Theory-Based Historical Surmise (THS) = Synthetic Experience (SE)

Synthetic Experience (SE) + Reflection on Synthetic Experience (RSE) = Improved Capacity for Judgment (ICJ)

These equations can be represented diagrammatically as shown below:

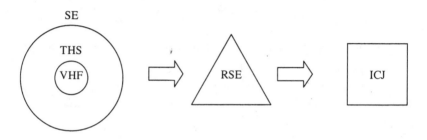

Appendix Two
Bach's *St. Matthew Passion* as a Model for Critical Analysis

"**P**ASSIONS" ARE A MUSICAL FORM in which biblical texts are used as the basis for the historical reenactment of the crucifixion of Jesus Christ, which is augmented by the composer's commentary on the event. They are part of a development in music of works devoted to reenactment, which are known as "historias." The purpose of performing passions is the promotion of active faith, which can be characterized as the exercise of a kind of intuitive knowledge. Listening to a performance can thus be thought of as a form of "critical analysis" with respect to the comprehension of religious experience.

The provenance of Clausewitz's conception of historical reenactment is not known. An intriguing possibility is that he found a model for his idea in Johann Sebastian Bach's *St. Matthew Passion*. In the 1820s, Felix Mendelssohn instigated efforts that resulted in performances of the *St. Matthew Passion* after more than a half-century of the work being neglected, if not unknown. These took place in Berlin, where Clausewitz resided, in 1829. Marie von Clausewitz's family, the von Brühls, had long associations with music in general and Bach in particular. Her grandfather had known Bach, and, indeed, had chosen his successor as cantor of St. Thomas's Church in Leipzig upon the great composer's death. Her cousin, Count Karl von Brühl, was the intendant of the Royal Opera and a member of the choral society (Singakademie) that performed the *St. Matthew Passion* in Berlin. The count knew Mendelssohn personally. Mendelssohn was also an undergraduate student studying under G. W. F. Hegel, the famous philosopher and an admirer of Bach's music. Hegel was an acquaintance of Clausewitz. It is at least possible, if not probable, that Clausewitz had, through his wife or her relations and his own social connections, knowledge of the intense discussion of the *St. Matthew Passion* that took place over the course of several years prior to its revival. And it seems likely that he attended the performance of the work in the spring of 1829.

Direct evidence to support the proposition that the *St. Matthew Passion* served as a model for Clausewitz's thinking about historical reenactment has yet to be discovered. A causal relationship between the two subjects may, indeed, not exist. But in any case, consideration of what Bach's great work and that of Clausewitz have in common with respect to historical method and instructive purpose can facilitate understanding of important qualities of the latter that might otherwise escape comprehension.

Notes

Preface and Acknowledgments

1. Michael Howard, "The Influence of Clausewitz," in Carl von Clausewitz, *On War*, trans. and ed. Michael Howard and Peter Paret (Princeton, NJ: Princeton University Press, 1976), pp. 29–30.

2. Julian S. Corbett, *Some Principles of Maritime Strategy* (Annapolis, MD: Naval Institute Press, 1998; first published 1911), p. 24.

3. Herbert Rosinski, *The German Army* (New York: Frederick A. Praeger, 1966), p. 110.

4. Bernard Brodie, "The Continuing Relevance of *On War*," in Clausewitz, *On War*, p. 53.

5. Peter Paret, "Clausewitz," in Peter Paret, ed., *Makers of Modern Strategy from Machiavelli to the Nuclear Age* (Princeton, NJ: Princeton University Press, 1986), p. 186.

6. Quoted in Hew Strachan, *Clausewitz's On War: A Biography* (New York: Atlantic Monthly Press, 2007), p. 2.

7. United States Army, Center of Military History, "The U.S. Army Chief of Staff's Professional Reading List," CMH Pub 105–5–1 (2006), Sublist 3.

8. See Bibliography.

9. Tiha von Ghyczy, Bolko von Oetinger, and Christopher Bassford, *Clausewitz on Strategy: Inspiration and Insight from a Master Strategist* (New York: John Wiley and Sons, 2001); Beatrice Heuser, *Reading Clausewitz* (London: Pimlico, 2002); Hugh Smith, *On Clausewitz: A Study of Military and Political Ideas* (New York: Palgrave Macmillan, 2005); Andreas Herberg-Rothe, *The Clausewitz Puzzle: The Political Theory of War* (Oxford: Oxford University Press, 2007; first published in German, 2001); Hew Strachan, *Clausewitz's On War* (2007).

10. See www.clausewitz.com/CWZHOME/CWZBASE.htm.

11. Clausewitz, *On War*, Book I, Chapter 1, p. 87.

12. The term "politics/policy" will henceforth replace the separate terms "politics" and "policy" in this book as a means of dealing with the dual meaning of the German word *Politik*. This convention will not be applied to quotations from the standard translation of *On War*.

13. Raymond Aron, *Clausewitz Philosopher of War*, trans. Christine Booker and Norman Stone (London: Routledge and Kegan Paul, 1983; first published in French, 1976), p. viii.

14. For example, see Michael I. Handel, *Masters of War: Classical Strategic Thought*, 3d ed., revised and expanded (London: Frank Cass, 2001), charts facing p. 350.

15. Bernard Brodie, "The Continuing Relevance of *On War*," in Clausewitz, *On War*, pp. 45–58, and Christopher Bassford, *Clausewitz in English: The Reception of Clausewitz in Britain and America, 1815–1945* (Oxford: Oxford University Press, 1994).

16. Azar Gat, *A History of Military Thought: From the Enlightenment to the Cold War* (Oxford: Oxford University Press, 2001 [1989]), pp. 254–255.

17. Bruce Fleming, "Can Reading Clausewitz Save Us from Future Mistakes?" *Parameters* 34 (Spring 2004): 76.

18. Tony Corn, "Clausewitz in Wonderland," *Policy Review,* September 2006, p. 3, http://www.hoover.org/publications/policyreview/4268401.html.

19. "Note of 10 July 1827," in Clausewitz, *On War,* p. 69 (italics in original).

20. *On War* is divided into eight parts, each one of which is called a "book." The books are divided into chapters.

21. "Unfinished Note, Presumably Written in 1830," in Clausewitz, *On War,* p. 70.

22. "Note of 10 July 1827," in Clausewitz, *On War,* pp. 69–70.

23. Gat, *History of Military Thought,* pp. 257–265.

24. Books I and II self-evidently constitute a consideration of the "nature of the whole." For the grouping of Books III, IV, and V, see Clausewitz, *On War,* III/1: 177. The connection between Books VI and VII is also obvious, but see Clausewitz, *On War,* VII/1: 523. For the concerns of Book VIII, see VIII/1: 577.

25. Clausewitz, *On War,* VIII/9: 633n.

26. Jon Sumida, "The Relationship between History and Theory in *On War:* The Clausewitzian Ideal and Its Implications," *Journal of Military History* 5 (April 2001), and "On Defense as the Stronger Form of War," Clausewitz in Hew Strachan and Andreas Herberg-Rothe, eds., *Clausewitz in the Twenty-First Century* (Oxford: Oxford University Press, 2007), pp. 163–181.

Introduction

1. Isaiah Berlin, "The Originality of Machiavelli" (originally published as "The Question of Machiavelli" in the *New York Review of Books,* 4 November 1971, pp. 20–31), in Isaiah Berlin, *The Proper Study of Mankind: An Anthology of Essays,* eds. Henry Hardy and Roger Hausheer (New York: Farrar, Straus and Giroux, 1998), p. 325.

2. For the works of Bassford, Heuser, and Strachan and Herberg-Rothe (Oxford University Clausewitz conference proceedings), see Bibliography. The Clausewitz "home page" is at www.clausewitz.com/CWZHOME/CWZBASE.htm.

3. W. B. Gallie, *Philosophers of Peace and War: Kant, Clausewitz, Marx, Engels, and Tolstoy* (Cambridge: Cambridge University Press, 1978), p. 49.

4. Stephen Jay Gould, "Men of the Thirty-Third Division: An Essay on Integrity," in *Eight Little Piggies: Reflections in Natural History* (New York: W. W. Norton, 1993), p. 125.

Chapter One. Theorists

1. "Note of 10 July 1827," in Carl von Clausewitz, *On War,* trans. and ed. Michael Howard and Peter Paret (Princeton, NJ: Princeton University Press, 1976), p. 70.

2. "On the Genesis of His Early Manuscript on the Theory of War, Written around 1818," Clausewitz, *On War,* p. 63.

3. "Note of 10 July 1827," p. 70.

4. Roger Parkinson, *Clausewitz* (New York: Stein and Day, 1971), p. 330. For Clausewitz's relationship to von der Groeben, see Peter Paret, *Clausewitz and the State* (Oxford: Clarendon, 1976), p. 258.

5. "Preface by Marie von Clausewitz to the Posthumous Edition of Her Husband's Works, Including *On War*," Clausewitz, *On War*, pp. 66–67.

6. John Shy, "Jomini," in Peter Paret, ed., *Makers of Modern Strategy from Machiavelli to the Nuclear Age* (Princeton, NJ: Princeton University Press, 1986), p. 153.

7. Captain John I. Alger, *Antoine-Henri Jomini: A Bibliographical Survey* (West Point, NY: United States Military Academy, 1975).

8. Baron de Jomini, *Summary of the Art of War, or, A New Analytical Compend of the Principal Combinations of Strategy, of Grand Tactics and of Military Policy*, trans. and ed. O. F. Winship and E. E. McLean (New York: G. P. Putnam, 1854) (henceforth cited as *Art of War* [Winship/McLean]), pp. 14–15.

9. Baron de Jomini, *Art of War*, trans. and ed. G. H. Mendell and W. P. Craighill (henceforth cited as *Art of War* [Mendell/Craighill]) (New York: Greenwood Press, 1971; first published Philadelphia: J. B. Lippincott, 1862).

10. See also VIII/9: 625.

11. See also Jomini, *Art of War* [Mendell/Craighill], pp. 37, 38, 116, 247, 250, 293–294, 296–297.

12. Clausewitz, *On War*, I/1: 84; I/2: 98–99; VI/1: 358; VI/5: 370; VII/2: 524; VII/5: 528; VII/15: 547; VII/22: 571–572.

13. Quoted in Major General J. F. C. Fuller, *A Military History of the Western World*, 3 vols. (New York: Minerva, 1967; first published 1955), II: 424.

14. Donald M. Schurman, *Julian S. Corbett, 1854–1922: Historian of British Maritime Policy from Drake to Jellicoe* (London: Royal Historical Society, 1981).

15. Julian S. Corbett, *Some Principles of Maritime Strategy*, introduction and notes by Eric J. Grove (Annapolis, MD: Naval Institute Press, 1988; first published 1911), p. 6n.

16. See also III/1: 181.

17. Preface, dated November 1903, in Julian S. Corbett, *England in the Mediterranean: A Study of the Rise and Influence of British Power within the Straits, 1603–1713*, 2d ed., 2 vols. (London: Longmans, Green, 1917; first published 1904), I: vii–viii.

18. Schurman, *Corbett*, p. 51.

19. Julian S. Corbett, "Strategical Terms and Definitions Used in Lectures on Naval History," in Corbett, *Some Principles of Maritime Strategy*, ed. Eric J. Grove (Annapolis, MD: Naval Institute Press, 1988; first published 1911), pp. 308–311.

20. Julian S. Corbett, "Notes on Strategy," in Corbett, *Some Principles*, pp. 327–332.

21. Julian S. Corbett, *England in the Seven Years' War: A Study in Combined Strategy*, 2d ed., 2 vols. (London: Longmans, Green, 1918; first published 1907), I: 24.

22. Corbett, *England in the Seven Years' War*, I: v–vi.

23. Clausewitz, *On War*, I/1: 80–81; VI/8: 387–388; VIII/3: 8.

24. Julian S. Corbett, *The Campaign of Trafalgar*, 2d ed., 2 vols. (London: Longmans, Green, 1919; first published 1909), II: 256–257. See also II: 301–302.

25. See also II: 283.

26. Nicholas A. Lambert, *Sir John Fisher's Naval Revolution* (Columbia: University of South Carolina Press, 1999).

27. Jon Sumida, "The Historian as Contemporary Analyst: Sir Julian Corbett and Admiral Sir John Fisher," in James Goldrick and John B. Hattendorf, eds., *Mahan Is Not Enough: The Proceedings of a Conference on the Works of Sir Julian Corbett and Admiral Sir Herbert Richmond* (Newport, RI: Naval War College Press, 1993), pp. 125–140.

28. See also "Unfinished Note, Presumably Written in 1830," in Clausewitz, *On War*, p. 71.

29. Brian Bond, *Liddell Hart: A Study of His Military Thought* (New Brunswick, NJ: Rutgers University Press, 1977); Alex Danchev, *Alchemist of War: The Life of Basil Liddell Hart* (London: Weidenfeld and Nicolson, 1998); Brian Holden Reid, "Hart, Sir Basil Henry Liddell," in Collin Matthew, H. C. G. Matthew, and Brian Howard Harrison, eds., *Oxford Dictionary of National Biography* (Oxford: Oxford University Press, 2004).

30. Bond, *Liddell Hart*, pp. 20–21; Danchev, *Alchemist of War*, p. 112.

31. Captain B. H. Liddell Hart, *Paris or the Future of War* (New York and London: Garland, 1972; first published 1925).

32. Captain B. H. Liddell Hart, "The Napoleonic Fallacy," in *The Remaking of Modern Armies* (London: John Murray, 1927).

33. Captain B. H. Liddell Hart, *A Greater Than Napoleon Scipio Africanus* (Edinburgh and London: William Blackwood and Sons, 1926), p. viii.

34. Captain B. H. Liddell Hart, *Great Captains Unveiled* (Edinburgh: W. Blackwood, 1927); *Reputations: Ten Years After* (Boston: Little, Brown, 1928); and *Sherman: Soldier, Realist, American* (New York: Dodd, Mead, 1929).

35. Spenser Wilkinson, "Killing No Murder: An Examination of Some New Theories of War," *Army Quarterly* 22 (October 1927): 14–27. See also Bond, *Liddell Hart*, pp. 50–51, and Christopher Bassford, *Clausewitz in English: The Reception of Clausewitz in Britain and America, 1815–1945* (New York: Oxford University Press, 1994), pp. 141–143.

36. B. H. Liddell Hart, *The Decisive Wars of History: A Study in Strategy* (Boston: Little, Brown, 1929).

37. Captain B. H. Liddell Hart, *The Real War, 1914–1918* (Boston: Little, Brown, 1930). Six years later, Liddell Hart would produce a shorter account of World War I that stated his negative views of Clausewitz at some length; see Liddell Hart, *The War in Outline* (New York: Random House, 1936), pp. 19, 49, 67–70.

38. Liddell Hart, *Foch: The Man of Orleans* (London: Eyre and Spottiswoode, 1931), p. 21. Language replicated in Liddell Hart, *The Ghost of Napoleon* (New Haven, CT: Yale University Press, 1934), p. 133.

39. See also pp. 20–21, 41, 55, 75.

40. For Liddell Hart as a "parable-artist," see Danchev, *Alchemist of War*, p. 176.

41. For later use of this passage, see Liddell Hart, *Strategy* (New York: Frederick A. Praeger, 1954), p. 356.

42. Bond, *Liddell Hart*, pp. 43–44, 51, 80–81; Bassford, *Clausewitz in English*, pp. 130–134; and Azar Gat, *A History of Military Thought: From the Enlightenment to the Cold War* (Oxford: Oxford University Press, 2001), pp. 692–693.

43. See also Liddell Hart, *Ghost of Napoleon*, p. 127.

44. For later use of this passage, see Liddell Hart, *Strategy*, p. 355.

45. Bond, *Liddell Hart*, p. 80.

46. For example, see his *Colonel Lawrence [of Arabia]: The Man behind the Legend*, new and enlarged ed. (New York: Halcyon House, 1935; first published 1934), pp. 30, 56, 127, 221; *War in Outline, 1914–1918*, pp. 19, 49, 67–70; *Europe in Arms* (New York: Random House, 1937), pp. 218–221, 228, 278; *Through the Fog of War* (New York: Random House, 1938), pp. 57–58; *The Defence of Britain* (New York: Random House, 1939), pp. 27–28, 36–37, 120, 189–190; *Thoughts on War* (London: Faber and Faber, 1944), pp. 33, 40, 43, 44, 58, 68, 126, 133–135, 158, 229, 295; *The Revolution in Warfare* (New Haven, CT: Yale University Press, 1947), pp. 38, 66–70, 76, 94; *Defence of the West* (New York: William Morrow, 1950), p. 314; *Strategy*, pp. 352–357; *The Liddell Hart Memoirs*, 2 vols. (New York: G. P. Putnam's Sons, 1965), I: 280.

47. For further discussion along these lines, see Clausewitz, *On War*, IV/3: 228; IV/4: 230; IV/9: 248, 251–252; IV/11: 258, 262; V/18: 354; VIII/6: 606.

48. For a restatement of the idea that an engagement offered and refused was tantamount to an engagement happening, see Clausewitz, *On War*, III/1: 181. For further descriptions of the character of war as an activity defined by fighting (which would include the prospect as well as the actuality of fighting), see Clausewitz, *On War*, II/1: 127; IV/3: 227.

49. See Sumida, "The Relationship of History and Theory." Although Liddell Hart did not make a direct attribution, this discussion was followed immediately by consideration of Clausewitz's views on the importance of the commander's will.

50. See also Liddell Hart, *Thoughts on War*, p. 22.

51. Liddell Hart, *The Ghost of Napoleon*, p. 126. Liddell Hart repeated this passage in works published in 1954 and 1967, for which see Liddell Hart, *Strategy*, p. 353, and Captain B. H. Liddell Hart, "Armed Forces and the Art of War: Armies," in J. P. T. Bury, ed., *The New Cambridge Modern History*, vol. 10, *The Zenith of European Power 1830–70* (Cambridge: Cambridge University Press, 1967), p. 318.

52. Liddell Hart, *Defence of Britain*, p. 28. In a radio interview of 1951, he relied on the authority of Clausewitz to justify his pre–World War II prediction that defensive action would overcome offensive, for which see Bassford, *Clausewitz in English*, p. 130.

53. Liddell Hart, *Decisive Wars*, p. 144. The same text is given in Liddell Hart, *Thoughts on War*, p. 270. See also Liddell Hart, *Thoughts on War*, p. 319.

54. Liddell Hart, *Thoughts on War*, p. 293. For Liddell Hart on the combination of defense and counterattack, see also *Europe in Arms*, p. 103; *The Revolution in Warfare*, p. 14; and *Strategy*, p. 328.

55. Danchev, *Alchemist of War*, p. 176; Jay Luvaas, "Clausewitz, Fuller and Liddell Hart," in Michael I. Handel, ed., *Clausewitz and Modern Strategy* (London: Frank Cass, 1986), p. 211.

56. Bassford, *Clausewitz in English*, p. 134.

57. Liddell Hart, *British Way of Warfare*, p. 109; *Thoughts on War*, p. 58.

58. See especially B. H. Liddell Hart, Foreword, in *Sun Tzu, The Art of War*, Samuel B. Griffith, trans. (London: Oxford University Press, 1963), p. vi.

59. B. H. Liddell Hart, *Why Don't We Learn from History?* (New York: Hawthorn Books, 1971; first published 1943), pp. 80–83.
60. Liddell Hart, *Defence of the West,* p. 245.
61. Danchev, *Alchemist of War,* p. 63.
62. Bassford, *Clausewitz in English,* pp. 132–133.

Chapter Two. Scholars

1. Carl von Clausewitz, *On War,* trans. and ed. Michael Howard and Peter Paret (Princeton, NJ: Princeton University Press, 1976).
2. Peter Paret, *Clausewitz and the State* (Oxford: Clarendon Press, 1976).
3. Raymond Aron, *Clausewitz: Philosopher of War,* trans. Christine Booker and Norman Stone (London: Routledge and Kegan Paul, 1983; first published in French, 1976).
4. W. B. Gallie, *Philosophers of Peace and War: Kant, Clausewitz, Marx, Engels and Tolstoy* (Cambridge: Cambridge University Press, 1978).
5. Raymond Aron, *Memoirs: Fifty Years of Political Reflection,* trans. and ed. George Holoch (New York: Holmes and Meier, 1990; first published in French, 1983).
6. For biographical information see Herbert Rosinski, *The Development of Naval Thought: Essays by Herbert Rosinski,* ed. B. Mitchell Simpson III (Newport, RI: Naval War College Press, 1977), p. vi, and Christopher Bassford, *Clausewitz in English: The Reception of Clausewitz in Britain and America, 1815–1945* (Oxford: Oxford University Press, 1994), pp. 186–189.
7. See also Aron, *Clausewitz,* p. 231.
8. See also Aron, *Clausewitz,* pp. 56–59, 87, 116.
9. See also Aron, *Clausewitz,* pp. 133, 209, 217.
10. See also Aron, *Clausewitz,* p. 89.
11. Roger Ashley Leonard, ed., *A Short Guide to Clausewitz,* On War (New York: G. P. Putnam's Sons, 1967), pp. 30–32.
12. Carl von Clausewitz, *On War,* ed. Anatol Rapoport (Harmondsworth: Penguin Books, 1968). For Aron's notice of the omission of Book VI in the Rapoport edition, see Aron, *Clausewitz,* p. 223.
13. Azar Gat, *A History of Military Thought* (Oxford: Oxford University Press, 2001 [1989]), pp. 257–265.
14. William H. Dray, *History as Re-Enactment: R. G. Collingwood's Idea of History* (Oxford: Clarendon Press, 1995).
15. Raymond Aron, *Introduction to the Philosophy of History: An Essay on the Limits of Historical Objectivity,* rev. ed., trans. George J. Irwin (Boston: Beacon Press, 1961; first published 1938; rev. ed. published 1948).
16. For example, see E. W. F. Tomlin, *R. G. Collingwood* (London: Longmans, Green, 1953), and Alan Donagan, *The Later Philosophy of R. G. Collingwood* (Oxford: Clarendon Press, 1962).
17. Aron, *Clausewitz,* pp. 234–238.

18. See Gat, *History of Military Thought*, p. 172.

19. Aron, *Clausewitz*, p. 3.

20. Bassford, *Clausewitz in English*, pp. 207–208.

21. See Peter Paret and John W. Shy, *Guerrillas in the 1960s* (London: Pall Mall, 1962), pp. 11–15; and the following works, all by Paret: "Clausewitz: A Bibliographical Survey," *World Politics* 17 (1965); "Clausewitz and the Nineteenth Century," in Michael Howard, ed., *The Theory and Practice of War* (London: Cassell, 1965); *Yorck and the Era of Prussian Reform, 1807–1815* (Princeton, NJ: Princeton University Press, 1966); "Clausewitz," in David L. Sills, ed., *International Encyclopedia of the Social Sciences*, vol. 2 (New York: Macmillan and Free Press, 1968), "Education, Politics, and War in the Life of Clausewitz," *Journal of the History of Ideas* 29 (1968); and *Clausewitz and the State* (Princeton, NJ: Princeton University Press, 1976).

22. Paret, *Clausewitz and the State*, pp. 6–10.

23. See also Paret, *Clausewitz and the State*, p. 294.

24. Paret, "The Genesis of *On War*," in Clausewitz, *On War*, eds. Howard and Paret.

25. Peter Paret, "Clausewitz," in Paret, ed., *Makers of Modern Strategy from Machiavelli to the Nuclear Age* (Princeton, NJ: Princeton University Press, 1986), pp. 186–213, and Peter Paret, "Introduction to Part One [Historical Writings]," in Carl von Clausewitz, *Historical and Political Writings*, trans. and ed. Peter Paret and Daniel Moran (Princeton, NJ: Princeton University Press, 1992), pp. 3–14.

26. Peter Paret, *Understanding War: Essays on Clausewitz and the History of Military Power* (Princeton, NJ: Princeton University Press, 1992), p. 130.

27. *Who Was Who, 1996–2000*, vol. 10 (New York: Palgrave, 2001), p. 203.

28. W. B. Gallie, *Peirce and Pragmatism* (New York: Dover, 1966; first published 1952), p. 40.

29. Gallie, *Peirce and Pragmatism*, p. 55.

30. W. B. Gallie, *Philosophy and the Historical Understanding* (London: Chatto and Windus, 1964), pp. 17–18.

31. Gallie, *Philosophers of Peace and War*, p. 1.

32. W. B. Gallie, "Clausewitz Today," *European Journal of Sociology* 19 (1978): 143–167. In the discussion that follows, I will refer to the printed version of the Wiles Lectures from *Philosophers of Peace and War*.

33. Captain B. H. Liddell Hart, *Strategy* (New York: Frederick A. Praeger, 1954), p. 352, and Major General J. F. C. Fuller, *The Conduct of War, 1789–1961: A Study of the Impact of the French, Industrial, and Russian Revolutions on War and Its Conduct* (Westport, CT: Greenwood, 1981; first published 1961), p. 60.

34. Material consistent with this interpretation can be found in other chapters. See Clausewitz, *On War*, I/2: 99; Gallie, *Philosophers of Peace and War*, note 23, p. 58. Gallie also noted that this view was expressed elsewhere. He did not offer specific references, but see Clausewitz, *On War*, I/1: 77, 88; III/16: 219; IV/10: 256–257; VI/30: 517–519; VIII/3: 592–593.

35. To be fair, while Paret did not emphasize or advertise, he did note (see the previous section of this chapter).

Chapter Three. Antecedents and Anticipations

1. Roger Parkinson, *Clausewitz: A Biography* (New York: Stein and Day, 1971), p. 35, and Peter Paret, *Clausewitz and the State* (Oxford: Clarendon, 1976), pp. 69, 81.

2. Peter Paret, *Clausewitz and the State* (Oxford: Clarendon Press, 1976), pp. 82–83, 169–208.

3. Azar Gat, *A History of Military Thought: From the Enlightenment to the Cold War* (Oxford: Oxford University Press, 2001), pp. 234–236; Paret, *Clausewitz and the State*, pp. 316–317.

4. Raymond Aron, *Clausewitz: Philosopher of War*, trans. Christine Booker and Norman Stone (London: Routledge and Kegan Paul, 1983; first published in French, 1976), 54, 226–232.

5. W. B. Gallie, *Philosophers of Peace and War: Kant, Clausewitz, Marx, Engels and Tolstoy* (Cambridge: Cambridge University Press, 1978), pp. 39, 41.

6. For the sources of this general account, see Bibliography.

7. Carl von Clausewitz, *Preussen in seiner grossen Katastrophe* (Vienna: Karolinger, 2001); "From *Observations on Prussia in Her Great Catastrophe*," in Carl von Clausewitz, *Historical and Political Writings*, trans. and ed. Peter Paret and Daniel Moran (Princeton, NJ: Princeton University Press, 1992), pp. 30–84.

8. Carl von Clausewitz, *Feldzug von 1812 in Russland*, in *Carl von Clausewitz: Schriften—Aufsätze—Studien—Briefe*, 2 vols. (vol. 2 in two parts), ed. Werner Hahlweg (Göttingen: Vandenhoeck and Ruprecht, 1990), II/ii: 717–935; Carl von Clausewitz, *The Campaign of 1812 in Russia* (New York: Da Capo Press, 1995; first published 1843); Carl von Clausewitz, "From *The Campaign of 1812 in Russia*," in Clausewitz, *Historical and Political Writings*, pp. 110–204; Carl von Clausewitz, *Der Feldzug von 1813 bis zum Waffenstillstand* (1813), unattributed free translation in General Count Gneisenau and J. E. Marston, *The Life and Campaigns of Field-Marshall Prince Blücher of Wahlstatt* (London: Sherwood, Neely, and Jones, 1815), pp. 50–148 (for the origin of this work, see Paret, *Clausewitz and the State*, p. 240); "The Strategic Critique of the Campaign of 1814 in France," in Clausewitz, *Historical and Political Writings*, pp. 205–219; Carl von Clausewitz, *Feldzug von 1815*, in *Schriften—Aufsätze—Studien—Briefe*, II/ii: 936–1118; Christopher Bassford and Gregory W. Pedlow, eds., *On Waterloo: The Exchange between Wellington and Clausewitz* (draft manuscript courtesy of Christopher Bassford); Parkinson, *Clausewitz*, pp. 125–127, 135; Paret, *Clausewitz and the State*, pp. 218–219; "Meine Vorlesungen über den kleinen Krieg, gehalten auf der Kriegs-Schule 1810 und 1811.—Artillerie, Geschütze," "Precis de la guerre en Espagne et en Portugal," and "Bekenntnisdenkschrift," in *Schriften—Aufsätze—Studien—Briefe*, I: 208–611, 708–750.

9. Paret, *Clausewitz and the State*, pp. 60–62; Charles Edward White, *The Enlightened Soldier: Scharnhorst and the Militärische Gesellschaft in Berlin, 1801–1805* (Westport, CT: Praeger, 1989), pp. 3–5.

10. Paret, *Clausewitz and the State*, p. 224.

11. Ibid., pp. 74–75.

12. Parkinson, *Clausewitz*, pp. 48, 56–57; Paret, *Clausewitz and the State*, p. 22.

13. Paret, *Clausewitz and the State*, pp. 147–148.

14. *Some Observations*, in Clausewitz, *Historical and Political Writings*, p. 77.

15. Clausewitz, *Campaign of 1812*, pp. 46, 52–53, 57, 60–61, 76, 80, 95–100, 110.

16. See also ibid., pp. 212–216.

17. See also ibid., p. 184. For Clausewitz's explanation of the significance of possessing a "European civilization," see Carl von Clausewitz, *On War*, trans. and ed. Michael Howard and Peter Paret (Princeton, NJ: Princeton University Press, 1976), VI/6: 375–376.

18. Paret, *Yorck*, pp. 155–156.

19. Ibid., p. 156. For the German troops in British service in Spain, see Sir Charles Oman, *Wellington's Army, 1809–1814* (London: Greenhill Books, 1986; first published 1913), pp. 221–225.

20. Don W. Alexander, *Rod of Iron: French Counterinsurgency Policy in Aragon during the Peninsular War* (Wilmington, DE: Scholarly Resources, 1985); John Lawrence Tone, *The Fatal Knot: The Guerrilla War in Navarre and the Defeat of Napoleon in Spain* (Chapel Hill: University of North Carolina Press, 1994); René Chartrand, "The Guerrillas: How Oman Underestimated the Role of Irregular Forces," in Paddy Griffith, ed., *A History of the Peninsular War*, vol. 9, *Modern Studies of the War in Spain and Portugal, 1808–1814* (London: Greenhill Books, 1999); and especially Charles J. Esdaile, *Fighting Napoleon: Guerrillas, Bandits and Adventurers in Spain, 1808–1814* (New Haven, CT: Yale University Press, 2004).

21. Parkinson, *Clausewitz*, p. 107.

22. Quoted in Parkinson, *Clausewitz*, p. 126. For the original German text, see Hahlweg, ed., *Schriften—Aufsätze—Studien—Briefe*, I: 231.

23. Hahlweg, ed., *Schriften—Aufsätze—Studien—Briefe*, I: 210–217.

24. Paret discussed Clausewitz's lectures about little wars at length, but did not explore the possibility that they were a Trojan horse for guerrilla war; see Paret, *Clausewitz and the State*, pp. 188–194. Parkinson and Werner Hahlweg do suggest this was the case; see Parkinson, *Clausewitz*, pp. 124–128, and Hahlweg, *Schriften—Aufsätze—Studien—Briefe*, I: 213. For a lecture text that suggests the resemblance of guerrilla war conducted by regular troops to guerrilla war waged by armed civilians, see Hahlweg, *Schriften—Aufsätze—Studien—Briefe*, I: 394. For the nominal association of Spanish guerrilla groups to specific regular military formations, see Chartrand, "The Guerrillas," p. 165. For the importance of regular troops in guerrilla war, see Esdaile, *Fighting Napoleon*, pp. 19, 21–22, 44–60, 105, 193–194. For the two most important studies of Clausewitz's views on People's War, see Werner Hahlweg, "Clausewitz and Guerrilla Warfare," in *Clausewitz and Modern Strategy*, edited by Michael I. Handel (Abingdon, UK: Frank Cass, 1986), 127–133; and Christopher Daase, "Clausewitz and Small Wars," in Hew Strachan and Andreas Herberg-Rothe, eds., *Clausewitz in the Twenty-First Century* (Oxford: Oxford University Press, 2007), pp. 182–195.

25. "Summary of the Instruction Given by the Author to His Royal Highness the Crown Prince in the Years 1810, 1811, and 1812," in Carl von Clausewitz, *On War*, trans. Colonel J. J. Graham, 3 vols. (London: Kegan Paul, Trench, Trübner, 1908), III: 211.

26. Parkinson, *Clausewitz*, p. 130. For Clausewitz's interest in the Peninsular War, see his "Precis de la guerre en Espagne et en Portugal," in Hahlweg, ed., *Schriften—Aufsätze—Studien—Briefe*, I: 599–611.

27. Parkinson, *Clausewitz*, p. 133.

28. For a translation of the first two parts of this three-part memorandum, see Clausewitz, *Historical and Political Writings*, pp. 285–303. Paret and Moran omitted the third part,

in which Clausewitz described his views on the military means of effective national resistance. See Hahlweg, ed., *Schriften—Aufsätze—Studien—Briefe,* I: 708–751. For a summary of the third part in English, see Parkinson, *Clausewitz,* pp. 134–135.

29. Clausewitz, "Bekenntnisdenkschrift," in Hahlweg, ed., *Schriften—Aufsätze—Studien—Briefe.*

30. Clausewitz, in Gneisenau/Marston, *Life of Blücher,* pp. 146–147.

31. Ibid., pp. 70–71, 126–127, 143–148.

32. Parkinson, *Clausewitz,* p. 225; Paret, *Clausewitz and the State,* p. 238.

33. Clausewitz, *On War* [Howard/Paret], VI/5: 370.

34. Clausewitz, *Historical and Political Writings,* p. 218.

35. Paret, *Clausewitz and the State,* p. 359.

36. Bassford and Pedlow, *On Waterloo,* Chapter 7. Carl von Clausewitz, *Feldzug von 1815,* in Hahlweg, ed., *Schriften—Aufsätze—Studien—Briefe,* II/ii: 936–938.

37. "Principles of the Art of War," in Clausewitz, *On War* [Graham], III: 182–183.

38. Ibid., III: 229.

39. Parkinson, *Clausewitz,* pp. 230–239; Paret, *Clausewitz and the State,* pp. 241–244.

40. Parkinson, *Clausewitz,* pp. 268–285; Paret, *Clausewitz and the State,* pp. 248–250.

41. See Parkinson, *Clausewitz,* pp. 97–130, 123, 141, 189–209, 214, 289–291; Paret, *Clausewitz and the State,* pp. 75, 229–230, 232–421; Clausewitz, *Campaign of 1812,* pp. 231–240; Paret, *Yorck,* pp. 191–196; Clausewitz-Gneisenau correspondence in Hahlweg, ed., *Schriften—Aufsätze—Studien—Briefe,* II/1.

42. Parkinson, *Clausewitz,* pp. 41–43, 145, 231; Paret, *Clausewitz and the State,* pp. 98–105, 107–109, 126, 193, 238. For Clausewitz's meeting with the tsar's sister, see Parkinson, *Clausewitz,* p. 179.

43. "On the Genesis of His Early Manuscript on the Theory of War, Written around 1818," in Carl von Clausewitz, *On War* [Howard/Paret], p. 63.

44. Ibid.

45. Clausewitz's target was undoubtedly a kind of reasoning characteristic of Jomini. See Baron de Jomini, *Art of War,* trans. and ed. G. H. Mendell and W. P. Craighill (New York: Greenwood Press, 1971; first published in Philadelphia: J. B. Lippincott, 1862), p. 65.

46. "Author's Preface to an Unpublished Manuscript on the Theory of War, Written between 1816 and 1818," in Clausewitz, *On War* [Howard/Paret], p. 61.

47. Ibid.

48. Gallie, *Philosophers of Peace and War,* p. 42.

49. W. B. Gallie, "Clausewitz Today," *European Journal of Sociology* 19 (1978): 144.

50. Ibid., p. 150.

51. Gallie, *Philosophers of Peace and War,* p. 46.

52. Quoted in Joseph Brent, *Charles Sanders Peirce: A Life* (Bloomington: Indiana University Press, 1993), p. 2.

53. Ibid., pp. 8–9.

54. Gallie, *Peirce and Pragmatism,* rev. ed. (New York: Dover, 1966; first published 1952). For Peirce's views on the power of chance to produce order and the relevance of this line of thought to chaos theory, see Brent, *Charles Sanders Peirce,* pp. 174–176. For the

possible relevance of this to Clausewitz, see Alan Beyerchen on nonlinearity in the next section of this chapter.

55. Ray Monk, *Ludwig Wittgenstein: The Duty of Genius* (New York: Free Press, 1990), and P. M. S. Hacker, "Wittgenstein, Ludwig Josef Johann," in H. C. G. Matthew and Brian Harrison, eds., *Oxford Dictionary of National Biography from the Earliest Times to the Year 2000* (Oxford: Oxford University Press, 2004), pp. 896–910. For a remarkably accessible introduction to Wittgenstein's formidable philosophical thought in the form of fiction, see Bruce Duffy, *The World as I Found It* (New York: Ticknor and Fields, 1987).

56. Gallie, *Peirce and Pragmatism,* p. 43.

57. Monk, *Duty of Genius,* pp. 215–224, 258–260, 273–274.

58. Ludwig Wittgenstein, *Philosophical Investigations,* 3d ed., trans. G. E. M. Anscombe (New York: Macmillan, 1971; first published 1953), p. 227 (italics in original).

59. Monk, *Duty of Genius,* pp. 530–531.

60. Quoted in Monk, *Duty of Genius,* p. 572.

61. E. W. F. Tomlin, *R. G. Collingwood* (London: Longmans, Green, for the British Council and the National Book League, 1953); T. M. Knox, "Collingwood, Robin George," in L. G. Wickham Legg and E. T. Williams, eds., *The Dictionary of National Biography, 1941–1950* (Oxford: Oxford University Press, 1959), pp. 168–170; and Peter Johnson, *R. G. Collingwood: An Introduction* (Bristol: Thoemmes Press, 1998).

62. "War," Collingwood wrote, "has been called a *continuation of policy*" (italics in original). R. G. Collingwood, *The New Leviathan, or Man, Society, Civilization and Barbarism,* rev. ed., ed. David Boucher (Oxford: Clarendon Press, 1992; first published 1942), p. 233.

63. William H. Dray, *History as Re-Enactment: R. G. Collingwood's Idea of History* (Oxford: Clarendon, 1995), p. 85, and Gary K. Browning, *Rethinking R. G. Collingwood: Philosophy, Politics and the Unity of Theory and Practice* (London: Palgrave Macmillan, 2004), pp. 102, 186n12.

64. Monk, *Duty of Genius,* p. 414.

65. R. G. Collingwood, *An Essay on Philosophical Method* (Bristol: Thoemmes Press, 1995; first published 1933), p. 204.

66. R. G. Collingwood, "The Nature and Aims of a Philosophy of History" (1924–1925), in William Debbins, ed., *Essays in the Philosophy of History* (Austin: University of Texas Press, 1965), pp. 34, 43. See also Dray, *History as Re-Enactment,* pp. 233–239, and R. G. Collingwood, *The Principles of History and Other Writings in Philosophy of History,* ed. W. H. Dray and W. J. van der Dussen (Oxford: Oxford University Press, 1999), pp. lxii–lxiii.

67. R. G. Collingwood, *An Autobiography* (Oxford: Oxford University Press, 1939).

68. R. G. Collingwood, *The Idea of History* (Oxford: Oxford University Press, 1956; first published 1946), p. 302.

69. Brent, *Charles Sanders Peirce,* p. 62.

70. Alan Beyerchen, "Clausewitz, Nonlinearity, and the Unpredictability of War," *International Security* 17 (Winter 1992/1993): 59–90.

71. See also Clausewitz, *On War* [Howard/Paret], I/3: 101, 102.

72. See also ibid., VIII/2: 580–581; VIII/3: 585–586.

73. Ibid., I/3: 102–103, 106, 108; III/1: 179; III/6: 191; III/7: 193; III/16: 217; and VIII/3: 586.

74. Guy Claxton, *Hare Brain, Tortoise Mind: How Intelligence Increases When You Think Less* (New York: Ecco, 1999; first published 1997), pp. 3–4.

75. Andrew Gordon to the author about a conversation with Claxton in the fall of 2004.

Chapter Four. Imagining High Command and Defining Strategic Choice

1. Carl von Clausewitz, *On War*, trans. and ed. Michael Howard and Peter Paret (Princeton, NJ: Princeton University Press, 1976), VII/1: 523.

2. See also ibid., VIII/2: 580.

3. See also ibid.

4. For the source of the concept in Enlightenment thought, see Azar Gat, *A History of Military Thought: From the Enlightenment to the Cold War* (Oxford: Oxford University Press, 2001), pp. 176–177.

5. For Clausewitz's further discussion of the conflict between the commander-in-chief and his subordinates, see IV/12: 264.

6. See also Clausewitz, *On War*, I/5: 116 (feelings act as higher judgment), and III/1: 177–178 (accurate fulfillment of unspoken assumptions only become evident in final success).

7. Note the anticipation of the discussion of political considerations and its effects on the attacker in Books VI and VII; see Clausewitz, *On War*, I/3: 112; VI/8: 387–388; VII/22: 573.

8. See also "Clausewitz's Final Notes Revisited," in Azar Gat, *A History of Military Thought from the Enlightenment to the Cold War* (Oxford: Oxford University Press, 2001), pp. 257–265.

9. Clausewitz, *On War*, VII/6: 529; VII/8: 533–534; VII/9: 535; VII/10: 536.

10. For anticipation of this discussion, see IV/12: 263. For its further development, see discussion of Book VII below.

11. See Peter Paret, *Clausewitz and the State* (Oxford: Clarendon Press, 1976), pp. 402–404; and Carl von Clausewitz, "Europe since the Polish Partitions" (1831) and "On the Basic Question of Germany's Existence" (1831), in *Historical and Political Writings*, trans. and ed. Peter Paret and Daniel Moran (Princeton, NJ: Princeton University Press, 1992), pp. 369–384.

12. Marie von Clausewitz's contention that her husband renounced all literary work after the spring of 1830 is contradicted by the fact that he wrote articles and memoranda afterward. For Marie von Clausewitz's remarks, see "Preface," *On War*, p. 66. For Clausewitz's writing in late 1830, see Roger Parkinson, *Clausewitz* (New York: Stein and Day, 1971), pp. 322–324, and Paret, *Clausewitz and the State*, pp. 402–409.

13. I am indebted to Terence Holmes for his persistent criticism of my original muddled treatment of Clausewitz's call for an offensive strategy against France.

Conclusions

1. Carl von Clausewitz, *On War*, trans. and ed. Peter Paret and Michael Howard (Princeton, NJ: Princeton University Press, 1976).

2. Ibid., III/11: 204; see also III/8: 194–197; V/3: 282–284.

3. Gallie virtually contradicts his contention that *On War* is a theory of practice in the conclusion to his book. See W. B. Gallie, *Philosophers of Peace and War: Kant, Clausewitz, Marx, Engels and Tolstoy* (Cambridge: Cambridge University Press, 1978), p. 135.

4. Michael Howard, *The Franco-Prussian War: The German Invasion of France, 1870–1871* (London: Methuen, 1981; first published 1961), p. 432.

5. For the availability throughout Europe of *The Essay on Man* in translation, see George S. Fraser, *Alexander Pope* (London: Routledge and Kegan Paul, 1978), pp. 5, 66, 68.

6. Alexander Pope, *An Essay on Man*, Epistle II (1733–1734), in Pat Rogers, ed., *Alexander Pope* (Oxford: Oxford University Press, 1993), p. 281.

7. "On the Genesis of His Early Manuscript on the Theory of War" (c. 1818), in Clausewitz, *On War*, p. 63.

8. Charles de Secondat Baron de Montesquieu, *The Spirit of Laws* (1748), trans. and ed. Anne M. Coler, Basia Carolyn Miller, and Harold Samuel Stone (Cambridge: Cambridge University Press, 1989), p. xliv.

9. Hermann Hesse, *Magister Ludi (The Glass Bead Game)*, trans. Richard and Clara Winston (New York: Bantam, 1970; first published in German, 1943), p. 151.

Bibliography

Clausewitz's Writing

On War

Graham, J. J., trans. Carl von Clausewitz, *On War*, 3 vols. London: Kegan Paul, Trench, Trübner, 1908.

———. Carl von Clausewitz, *On War*. Introduction by Jan Willem Honig. New York: Barnes and Noble, 2004.

Graham, J. J., trans., and Anatol Rapoport, ed. Carl von Clausewitz, *On War*, abridged. Harmondsworth, UK: Penguin, 1968.

Greene, Joseph I., ed. *The Essential Clausewitz: Selections from* On War. Mineola, NY: Dover, 2003; first published 1945.

Hahlweg, Werner, ed. Carl von Clausewitz, *Vom Kriege*. Bonn: Ferdinand Dümmlers Verlag, 1980.

Howard, Michael, and Peter Paret, trans. and ed. Carl von Clausewitz, *On War*. Princeton, NJ: Princeton University Press, 1976.

———. Carl von Clausewitz, *On War*. Everyman's Library, vol. 121. New York: Alfred A. Knopf, 1993.

Howard, Michael, Peter Paret, and Beatrice Heuser, eds. Carl von Clausewitz, *On War*, abridged. Oxford: Oxford University Press, 2007.

Jolles, O. J. Matthijs, trans. Karl von Clausewitz, *On War*. New York: Modern Library, 1943.

Other Works

Bassford, Christopher, and Gregory W. Pedlow, eds. *On Waterloo: The Exchange between Wellington and Clausewitz*. Draft manuscript courtesy of Christopher Bassford.

Clausewitz, Carl von. *Ausgewählte Briefe an Marie von Clausewitz und Gneisenau*. Berlin: Verlag der Nation, 1953.

———. *The Campaign of 1812 in Russia*. New York: Da Capo Press, 1995.

Gatzke, Hans W. *Principles of War*. Mineola, NY: Dover, 2003; first published 1942.

Gneisenau, General Count, and J. E. Marston. "Free Translation of Carl von Clausewitz, *Der Feldzug von 1813 bis zum Waffenstillstand*," in *The Life and Campaigns of Field-Marschall Prince Blücher of Wahlstaatt*. London: Sherwood, Neely, and Jones, 1815.

Hahlweg, Werner, ed. *Carl von Clausewitz: Schriften—Aufsätze—Studien—Briefe*, 2 vols. (vol. 2, in two parts). Göttingen: Vandenhoeck und Ruprecht, 1966–1990.

Niemeyer, Joachim, ed. *Carl von Clausewitz: Historische Briefe über die grossen Kriegsereignisse im Oktober 1806*. Bonn: Ferdinand Dümmlers Verlag, 1977.

Ungelter, Peter, ed. Carl von Clausewitz, *Preussen in seiner grossen Katastrophe*. Wien: Karolinger, 2001.

Bibliography

Biographical Studies

Hahlweg, Werner. *Clausewitz: Soldat, Politiker, Denker.* Göttingen: Musterschmidt, 1969.

Howard, Michael. *Clausewitz.* Oxford: Oxford University Press, 1983.

Parkinson, Roger. *Clausewitz: A Biography.* New York: Stein and Day, 1971.

Schramm, Wilhelm von. *Clausewitz: General und Philosoph.* Munich: Wilhelm Heyne, 1982; first published 1976.

Schwartz, Karl. *Leben des Generals Carl von Clausewitz und der Frau Marie von Clausewitz geb. Gräfin von Brühl,* 2 vols. Berlin: Ferdinand Dümmlers Verlag, 1878.

Critical Studies

Note: See section entitled "Theorists, Scholars, and Philosophers" for works by or about Jomini, Corbett, Liddell Hart, Aron, Paret, Gallie, Peirce, Wittgenstein, and Collingwood.

Bassford, Christopher. "Jomini and Clausewitz: Their Interaction." Consortium on Revolutionary Europe, Georgia State University, 26 February 1993, http://www.clausewitz.com/CWZHOME/Jomini/JOMINIX.htm.

———. *Clausewitz in English: The Reception of Clausewitz in Britain and America, 1815–1945.* Oxford: Oxford University Press, 1994.

———. "*On War* 2000: A Research Proposal," 5 October 2006, http://www.clausewitz.com/CWZHOME/complex/Proposax.htm.

Beyerchen, Alan. "Clausewitz, Nonlinearity, and the Unpredictability of War." *International Security* 17 (Winter 1992/1993): 59–90.

Brodie, Bernard. "On Clausewitz: A Passion for War." *World Politics* 25 (January 1973): 288–308.

Corn, Tony. "Clausewitz in Wonderland." *Real Clear Politics,* 9 September 2006, http://www.realclearpolitics.com/articles/2006/09/clausewitz_in_wonderland.html, and http://www.policyreview.org/000/corn2.html.

Echevarria, Antuilio J. *After Clausewitz: German Military Thinkers before the Great War.* Lawrence: University Press of Kansas, 2000.

———. "Clausewitz's Center of Gravity: It's *Not* What We Thought." *Naval War College Review* 56 (Winter 2003): 108–123.

———. "Principles of War or Principles of Battle?" in Anthony D. McIvor, ed., *Rethinking the Principles of War.* Annapolis, MD: Naval Institute Press, 2005.

Fleming, Bruce. "Can Reading Clausewitz Save Us from Future Mistakes?" *Parameters* 34 (Spring 2004): 62–76.

Gat, Azar. *A History of Military Thought: From the Enlightenment to the Cold War.* Oxford: Oxford University Press, 2001.

Ghyczy, Tiha von, Bolko von Oetinger, and Christopher Bassford. *Clausewitz on Strategy: Inspiration and Insight from a Master Strategist.* New York: John Wiley, 2001.

Hahlweg, Werner. *Lehrmeister des kleinen Krieges von Clausewitz bis Mao Tse-tung und Che Guevara.* Darmstadt: Wehr und Wissen, 1968.

Bibliography

Handel, Michael I., ed. *Clausewitz and Modern Strategy*. Abingdon, UK: Frank Cass, 1986.

———. *Masters of War: Sun Tzu, Clausewitz and Jomini*. London: Frank Cass, 1992.

———. *Masters of War: Classical Strategic Thought*, 3d ed., revised and expanded. London: Frank Cass, 2001.

Herberg-Rothe, Andreas. *Das Rätsel Clausewitz: Politische Theorie des Krieges im Widerstreit*. Munich: Wilhelm Fink, 2001. Published in English as *Clausewitz's Puzzle: The Political Theory of War*. Oxford: Oxford University Press, 2007.

Heuser, Beatrice. *Reading Clausewitz*. London: Pimlico, 2002.

Holmes, Terrence M. "Planning versus Chaos in Clausewitz's *On War*." *Journal of Strategic Studies* 30 (February 2007): 129–151.

Huber, Commander Jeff. "Clausewitz Is Dead." *United States Naval Institute Proceedings* 127 (March 2001): 119–120.

Kiesling, Eugenia C. "*On War* without the Fog." *Military Review* 81 (September-October 2001): 85–87.

Leonard, Roger Ashley. *A Short Guide to Clausewitz*, On War. New York: G. P. Putnam's Sons, 1967.

Otte, T. G. "Educating Bellona: Carl von Clausewitz and Military Education," in Gregory C. Kennedy and Keith Neilson, eds., *Military Education: Past, Present, and Future*. Westport, CT: Praeger, 2002.

Porch, Douglas. "Writing History in the 'End of History' Era—Reflections on Historians and the GWOT." *Journal of Military History* 70 (October 2006): 1065–1080.

Rothfels, H. "Clausewitz," in Edward Mead Earle, ed., *Makers of Modern Strategy: Military Thought from Machiavelli to Hitler*. Princeton, NJ: Princeton University Press, 1941.

Schering, Walther Malmsten. *Carl von Clausewitz. Geist und Tat: Das Vermächtnis des Soldaten und Denkers*. Stuttgart: Alfred Kröner, 1941.

Smith, Hugh. *On Clausewitz: A Study of Military and Political Ideas*. New York: Palgrave Macmillan, 2005.

Smith, M. L. R. "Strategy in an Age of 'Low-Intensity' Warfare: Why Clausewitz Is Still More Relevant Than His Critics," in Isabelle Duyvesteyn and Jan Angstrom, eds., *Rethinking the Nature of War*. London: Frank Cass, 2005.

Strachan, Hew. *Clausewitz's On War: A Biography*. New York: Atlantic Monthly Press, 2007.

Strachan, Hew, and Andreas Herthberg-Rothe, eds. *Clausewitz in the Twenty-First Century*. Oxford: Oxford University Press, 2007.

Sumida, Jon Tetsuro. *Inventing Grand Strategy and Teaching Command: The Classic Works of Alfred Thayer Mahan Reconsidered*. Washington, DC: Woodrow Wilson Center Press, and Baltimore: Johns Hopkins University Press, 1997.

———. "History and Theory: The Clausewitzian Ideal and Its Implications." *Journal of the Royal United Services Institute of Australia* 21 (May 2000): 63–73.

———. "History and Theory: The Clausewitzian Ideal and Its Implications," in David Stevens and John Reeve, eds., *Southern Trident: Strategy, History and the Rise of Australian Naval Power*. Crows Nest, Sydney, Australia: Allen and Unwin, 2001.

———. "The Relationship between History and Theory in *On War*: The Clausewitzian Ideal and Its Implications." *Journal of Military History* 65 (April 2001): 333–354.

Bibliography

——. "Response by Jon Sumida to the Remarks by Dr. Rogers." *Journal of Military History* 66 (October 2002): 1176.

——. "On Defense as the Stronger Form of War." Clausewitz in the 21st Century Conference, Department of Politics and International Relations, 21–23 March 2005, http://ccw.politics.ox.ac.uk/events/archives/tt05_clausewitz_sumida.pdf. Also in Hew Strachan and Andreas Herthberg-Rothe, eds., *Clausewitz in the Twenty-First Century*. Oxford: Oxford University Press, 2007.

——. "Pitfalls and Prospects: The Misuses and Uses of Military History and Classical Theory in an Age of Transformation," in Anthony McIvor, ed., *Rethinking the Principles of War*. Annapolis, MD: Naval Institute Press, 2005.

Summers, Harry G. *On Strategy: A Critical Analysis of the Vietnam War*. Novato, CA: Presidio, 1982.

United States Military Academy, Department of Military Art and Engineering. *Jomini, Clausewitz and Schlieffen*. West Point, NY: 1942.

Van Creveld, Martin. *Command in War*. Cambridge, MA: Harvard University Press, 1985.

——. *The Art of War: War and Military Thought*. London: Cassell, 2000.

Watts, Barry D. *Clausewitzian Friction and Future War*. Washington, DC: Institute for National Strategic Studies, National Defense University, 1996.

Historical Studies of Prussia and the Wars of the French Revolution and Empire, and Related Subjects

Alexander, Don W. *Rod of Iron: French Counterinsurgency Policy in Aragon during the Peninsular War*. Wilmington, DE: Scholarly Resources, 1985.

Best, Geoffrey. *War and Society in Revolutionary Europe, 1770–1870*. Oxford: Oxford University Press, 1986.

Chalfont, Lord, ed. *Waterloo: Battle of Three Armies*. New York: Alfred A. Knopf, 1980.

Chandler, David G. *The Campaigns of Napoleon*. New York: Macmillan, 1966.

——. *Dictionary of the Napoleonic Wars*. New York: Macmillan, 1979.

——. *Jena 1806: Napoleon Destroys Prussia*. London: Osprey, 1993.

——. *On the Napoleonic Wars: Collected Essays*. London: Greenhill Books, 1994.

Chartrand, René. "The Guerrillas: How Oman Underestimated the Role of Irregular Forces," in Paddy Griffith, ed., *A History of the Peninsular War*. Vol. 9, *Modern Studies of the War in Spain and Portugal, 1808–1814*. London: Greenhill Books, 1999.

Craig, Gordon. *The Politics of the Prussian Army, 1640–1945*. Oxford: Clarendon Press, 1955.

Esdaile, Charles J. *The Peninsular War: A New History*. Basingstoke, UK: Palgrave Macmillan, 2003.

——. *Fighting Napoleon: Guerrillas, Bandits and Adventurers in Spain, 1808–1814*. New Haven, CT: Yale University Press, 2004.

Fuller, J. F. C. *A Military History of the Western World*, 3 vols. New York: Minerva, 1967; first published 1955.

———. *The Conduct of War, 1789–1961: A Study of the Impact of the French, Industrial, and Russian Revolutions on War and Its Conduct*. Westport, CT: Greenwood, 1981; first published 1961.

Fussell, Paul. *The Great War and Modern Memory*. London: Oxford University Press, 1975.

Gooch, John. *Armies in Europe*. London: Routledge and Kegan Paul, 1980.

Hall, Christopher D. *British Strategy in the Napoleonic War, 1803–15*. Manchester: Manchester University Press, 1992.

Hofschröer, Peter. *Prussian Reserve, Militia & Irregular Troops, 1806–15*. London: Osprey, 1987.

———. *1815: The Waterloo Campaign. Wellington, His German Allies and the Battles of Ligny and Quatre Bras*. London: Greenhill Books, 1998.

———. *1815: The Waterloo Campaign. The German Victory*. London: Greenhill Books, 1999.

———. *Lützen & Bautzen 1813: The Turning Point*. London: Osprey, 2001.

Holborn, Hajo. *A History of Modern Germany, 1648–1840*. New York: Alfred A. Knopf, 1969.

Howard, Michael. *The Franco-Prussian War: The German Invasion of France, 1870–1871*. London: Rupert Hart-Davis, 1961.

———. *Studies in War and Peace*. New York: Viking, 1972.

Kaiser, David. *Politics and War: European Conflict from Philip II to Hitler*. Cambridge, MA: Harvard University Press, 1990.

Leggiere, Michael V. *Napoleon and Berlin: The Franco-Prussian War in North Germany, 1813*. Norman: University of Oklahoma Press, 2002.

———. "Napoleon's Gamble in North Germany." *Journal of Military History* 67 (January 2003): 39–84.

Maude, F. N. *The Jena Campaign: 1806: The Twin Battles of Jena and Auerstadt between Napoleon's French and the Prussian Army*. UK: Leonaur, 2007; first published in 1909.

Meinecke, Friedrich. *The Age of German Liberation, 1795–1815*. Translated by Peter Paret and Helmuth Fischer. Berkeley: University of California Press, 1977; first published in German, 1957.

Moran, Daniel. "Arms and the Concert: The Nation in Arms and the Dilemmas of German Liberalism," in Daniel Moran and Arthur Waldron, eds., *The People in Arms: Military Myth and National Mobilization since the French Revolution*. Cambridge: Cambridge University Press, 2003.

Nafziger, George F. *Napoleon's Invasion of Russia*. Novato, CA: Presidio Press, 1988.

———. *Lutzen [sic] & Bautzen: Napoleon's Spring Campaign of 1813*. Chicago: Emperor's Press, 1992.

———. *Napoleon at Dresden: The Battles of August 1813*. Chicago: Emperor's Press, 1994.

———. *Napoleon at Leipzig: The Battle of Nations, 1813*. Chicago: Emperor's Press, 1996.

Oman, Sir Charles. *Wellington's Army, 1809–1814*. London: Greenhill Books, 1986; first published 1913.

Paret, Peter. *Yorck and the Era of Prussian Reform, 1807–1815*. Princeton, NJ: Princeton University Press, 1966.

Petre, F. Loraine. *Napoleon's Conquest of Prussia: 1806*. London: Arms and Armour Press, 1972; first published 1907.

Bibliography

——. *Napoleon's Campaign in Poland, 1806–1807,* 3d ed. Barton-under-Needwood, UK: Wren's Park, 2001; first published 1907.

——. *Napoleon's Last Campaign in Germany, 1813.* London: Greenhill Books, 1992; first published 1912.

——. *Napoleon at Bay, 1814.* London: Arms and Armour Press, 1977; first published 1914.

Ritter, Gerhard. *The Sword and the Scepter: The Problem of Militarism in Germany.* Vol. 1, *The Prussian Tradition, 1740–1890,* 3d ed. Translated by Heinz Norden. Coral Gables, FL: University of Miami Press, 1969.

Rosinski, Herbert. *The German Army.* New York: Frederick A. Praeger, 1966.

——. *The Development of Naval Thought: Essays by Herbert Rosinski.* Edited by B. Mitchell Simpson III. Newport, RI: Naval War College Press, 1977.

Simms, Brendan. *The Impact of Napoleon: Prussian High Politics, Foreign Policy and the Crisis of the Executive, 1797–1806.* Cambridge: Cambridge University Press, 1997.

Strachan, Hew. *European Armies and the Conduct of War.* London: George Allen and Unwin, 1983.

Sutherland, D. M. G. *France, 1789–1815: Revolution and Counterrevolution.* London: Fontana/Collins, 1985.

Tone, John Lawrence. *The Fatal Knot: The Guerrilla War in Navarre and the Defeat of Napoleon in Spain.* Chapel Hill: University of North Carolina Press, 1994.

Wawro, Geoffrey. *The Franco-Prussian War: The German Conquest of France in 1870–1871.* Cambridge: Cambridge University Press, 2003.

Weigley, Russell F. *The Age of Battles: The Quest for Decisive Warfare from Breitenfeld to Waterloo.* Bloomington: Indiana University Press, 1991.

White, Charles Edward. *The Enlightened Soldier: Scharnhorst and the Militärische Gesellschaft in Berlin, 1801–1805.* Westport, CT: Praeger, 1989.

Theorists, Scholars, and Philosophers

Berlin, Isaiah. *The Proper Study of Mankind: An Anthology of Essays.* Edited by Henry Hardy and Roger Hausheer. New York: Farrar, Straus and Giroux, 1998.

Breisach, Ernst. *Historiography: Ancient, Medieval, and Modern.* Chicago: University of Chicago Press, 1983.

Gould, Stephen Jay. *Eight Little Piggies: Reflections in Natural History.* New York: W. W. Norton, 1993.

Hegel, Georg Wilhelm Friedrich. *The Philosophy of History.* Translated by J. Sibree. Edited by Charles Hegel. New York: Dover, 1956; first published 1899.

Hesse, Hermann. *Magister Ludi (The Glass Bead Game).* Translated by Richard and Clara Winston. New York: Bantam, 1970; first published in German, 1943.

——. *My Belief: Essays on Life and Art.* Translated by Denver Lindley. Edited by Theodore Ziolkowski. New York: Farrar, Straus and Giroux, 1974.

James, William. *The Varieties of Religious Experience: A Study in Human Nature.* New York: Mentor, 1958; first published 1902.

Bibliography

Krieger, Leonard. *The German Idea of Freedom: History of a Political Tradition.* Chicago: University of Chicago Press, 1957.

Macmurray, John. *The Self as Agent.* New York: Humanity, 1991; first published 1957.

Meinecke, Friedrich. *Historism: The Rise of a New Historical Outlook.* Translated by J. E. Anderson. New York: Herder and Herder, 1972; first published in German, 1959.

Montesquieu, Baron de [Charles de Secondat]. *The Spirit of Laws.* Translated and edited by Anne M. Cohler, Basia Carolyn Miller, and Harold Samuel Stone. Cambridge: Cambridge University Press, 1989.

Pinkard, Terry. *Hegel: A Biography.* Cambridge: Cambridge University Press, 2000.

Jomini

Alger, Captain John I. *Antoine-Henri Jomini: A Bibliographical Survey.* West Point, NY: United States Military Academy, 1975.

Brinton, Crane, Gordon A. Craig, and Felix Gilbert. "Jomini," in Edward Mead Earle, ed., *Makers of Modern Strategy: Military Thought from Machiavelli to Hitler.* Princeton, NJ: Princeton University Press, 1941.

Jomini, Baron de. *Art of War.* Translated and edited by G. H. Mendell and W. P. Craighill. New York: Greenwood Press, 1971; first published 1862; first published in French, 1838.

———. *Summary of the Art of War, or, A New Analytical Compend of the Principal Combinations of Strategy, of Grand Tactics and of Military Policy.* Translated and edited by O. F. Winship and E. E. McLean. New York: G. P. Putnam, 1854; first published in French, 1838.

Shy, John. "Jomini," in Peter Paret, ed., *Makers of Modern Strategy from Machiavelli to the Nuclear Age.* Princeton, NJ: Princeton University Press, 1986.

Corbett

Corbett, Julian S. *England in the Mediterranean: A Study of the Rise and Influence of British Power within the Straits, 1603–1713,* 2 vols., 2d ed. London: Longmans, Green, 1917; first published 1904.

———. *England in the Seven Years' War: A Study in Combined Strategy,* 2 vols., 2d ed. London: Longmans, Green, 1918; first published 1907.

———. *The Campaign of Trafalgar,* 2 vols., 2d ed. London: Longmans, Green, 1919; first published 1909.

———. *Some Principles of Maritime Strategy.* Annapolis, MD: Naval Institute Press, 1988; first published 1911.

Lambert, Nicholas. *Sir John Fisher's Naval Revolution.* Columbia: University of South Carolina Press, 1999.

Schurman, Donald M. *Julian S. Corbett, 1854–1922: Historian of British Maritime Policy from Drake to Jellicoe.* London: Royal Historical Society, 1981.

Sumida, Jon Tetsuro. *In Defence of Naval Supremacy: Finance, Technology and British Naval Policy, 1889–1914.* Boston: Unwin Hyman, 1989.

———. "The Historian as Contemporary Analyst: Sir Julian Corbett and Admiral Sir John Fisher," in James Goldrick and John B. Hattendorf, eds., *Mahan Is Not Enough: The*

Bibliography

Proceedings of a Conference on the Works of Sir Julian Corbett and Admiral Sir Herbert Richmond. Newport, RI: Naval War College Press, 1993.

Wilkinson, Spenser. "Strategy in the Navy." *Morning Post,* 3 August 1909, http://clausewitz.com/CWZHOME/HISTART/WILK1.html.

———. "Strategy at Sea." *Morning Post,* 19 February 1912, http://clausewitz.com/CWZHOME/HISTART/WILK1.html.

Liddell Hart

Bond, Brian. *Liddell Hart: A Study of His Military Thought.* New Brunswick, NJ: Rutgers University Press, 1977.

Danchev, Alex. *Alchemist of War: The Life of Basil Liddell Hart.* London: Weidenfeld and Nicolson, 1988.

Liddell Hart, Captain B. H. *Paris, Or the Future of War.* New York and London: Garland, 1972; first published 1925.

———. *A Greater Than Napoleon Scipio Africanus.* Edinburgh and London: William Blackwood and Sons, 1926.

———. *Great Captains Unveiled.* Edinburgh: W. Blackwood, 1927.

———. *The Remaking of Modern Armies.* London: John Murray, 1927.

———. *Reputations: Ten Years After.* Boston: Little, Brown, 1928.

———. *The Decisive Wars of History: A Study in Strategy.* Boston: Little, Brown, 1929.

———. *Sherman: Soldier, Realist, American.* New York: Dodd, Mead, 1929.

———. *The Real War, 1914–1918.* Boston: Little, Brown, 1930.

———. *The British Way in Warfare.* London: Faber and Faber, 1931.

———. *Foch: The Man of Orleans.* London: Eyre and Spottiswoode, 1931.

———. *Colonel Lawrence [of Arabia]: The Man behind the Legend,* new and enlarged ed. New York: Halcyon House, 1935; first published 1934.

———. *The Ghost of Napoleon.* New Haven, CT: Yale University Press, 1934.

———. *The War in Outline.* New York: Random House, 1936.

———. *The Defence of Britain.* New York: Random House, 1939.

———. *Why Don't We Learn from History?* New York: Hawthorn Books, 1971; first published 1943.

———. *Thoughts on War.* London: Faber and Faber, 1944.

———. *The Revolution in Warfare.* New Haven, CT: Yale University Press, 1947.

———. *Defence of the West.* New York: William Morrow, 1950.

———. *Strategy.* New York: Frederick A. Praeger, 1954.

———. "Foreword," in Sun Tzu, *The Art of War.* Translated by Samuel B. Griffith. London: Oxford University Press, 1963.

———. *The Liddell Hart Memoirs,* 2 vols. New York: G. P. Putnam's Sons, 1965.

———. "Armed Forces and the Art of War: Armies," in J. P. T. Bury, ed., *The New Cambridge Modern History.* Vol. 10, *The Zenith of European Power, 1830–70.* Cambridge: Cambridge University Press, 1967.

Mearsheimer, John J. *Liddell Hart and the Weight of History.* Ithaca, NY: Cornell University Press, 1988.

Reid, Brian Holden. "Hart, Sir Basil Liddell," in Collin Matthew, H. C. G. Matthew, and Brian Howard Harrison, eds., *Oxford Dictionary of National Biography*. Oxford: Oxford University Press, 2004.

Wilkinson, Spenser. "Killing No Murder: An Examination of Some New Theories of War." *Army Quarterly* 22 (October 1927): 14–27.

Aron

Anderson, Brian C. *Raymond Aron: The Recovery of the Political*. Lanham, MD: Rowman and Littlefield, 1997.

Aron, Raymond. *Introduction to the Philosophy of History: An Essay on the Limits of Historical Objectivity*, rev. ed. Translated by George J. Irwin. Boston: Beacon Press, 1961; first published in French, 1938; rev. ed. published in French, 1948.

———. *Peace and War: A Theory of International Relations*. Translated by Richard Howard and Annette Baker Fox. New York: Frederick A. Praeger, 1968; first published in French, 1962.

———. *Clausewitz: Philosopher of War*. Translated by Christine Booker and Norman Stone. London: Routledge and Kegan Paul, 1983; first published in French, 1976.

———. *Politics and History: Selected Essays by Raymond Aron*. Translated and edited by Miriam Bernheim Conant. New York: Free Press, 1978.

———. *Memoirs: Fifty Years of Political Reflection*, abridged ed. Translated and edited by George Holoch. New York: Holmes and Meier, 1990; first published in French, 1983.

Mahoney, Daniel J. *The Liberal Political Science of Raymond Aron: A Critical Introduction*. Lanham, MD: Rowman and Littlefield, 1992.

Paret

Brodie, Bernard. "In Quest of the Unknown Clausewitz: A Review. *Clausewitz and the State* by Peter Paret." *International Security* 1 (1977): 62–69.

Clausewitz, Carl von. *Historical and Political Writings*. Translated and edited by Peter Paret and Daniel Moran. Princeton, NJ: Princeton University Press, 1992.

Paret, Peter. "Clausewitz: A Bibliographical Survey." *World Politics* 17 (1965).

———. "Clausewitz and the Nineteenth Century," in Michael Howard, ed., *The Theory and Practice of War*. London: Cassell, 1965.

———. *Yorck and the Era of Prussian Reform, 1807–1815*. Princeton, NJ: Princeton University Press, 1966.

———. "Clausewitz," in David L. Sills, ed., *International Encyclopedia of the Social Sciences*, vol. 2. New York: Macmillan and Free Press, 1968.

———. "Education, Politics, and War in the Life of Clauswitz." *Journal of the History of Ideas* 29 (1968).

———. "Nationalism and the Sense of Military Obligation." *Military Affairs* 34 (February 1970): 2–6.

———. *Clausewitz and the State*. Oxford: Clarendon Press, 1976.

———. Review [untitled] of *Penser la Guerre, Clausewitz* by Raymond Aron. *Journal of Interdisciplinary History* 8 (Autumn 1977): 369–372.

Bibliography

———. Review [untitled] of *Philosophers of Peace and War* by W. B. Gallie. *Journal of Modern History* 51 (March 1979): 114–116.

———. "Clausewitz," in Peter Paret, ed., *Makers of Modern Strategy: From Machiavelli to the Nuclear Age.* Princeton, NJ: Princeton University Press, 1986.

———. Review [untitled] of *Clausewitz: Philosopher of War* by Raymond Aron. *Military Affairs* 50 (July 1986): 159–160.

———. "An Unknown Letter by Clausewitz." *Journal of Military History* 55 (April 1991): 145–151.

———. *Understanding War: Essays on Clausewitz and the History of Military Power.* Princeton, NJ: Princeton University Press, 1992.

———. *Clausewitz and the State: The Man, His Theories, and His Times.* Reissue with a new Preface. Princeton, NJ: Princeton University Press, 2007.

Paret, Peter, and John W. Shy. *Guerrillas in the 1960s.* London: Pall Mall, 1962.

Gallie

Gallie, W. B. *Peirce and Pragmatism.* New York: Dover, 1966; first published 1952.

———. "Peirce's Pragmaticism," in Philip P. Wiener and Frederic H. Young, eds., *Studies in the Philosophy of Charles Sanders Peirce.* Cambridge, MA: Harvard University Press, 1952.

———. *Philosophy and the Historical Understanding.* London: Chatto and Windus, 1964.

———. "Clausewitz Today." *European Journal of Sociology* 19 (1978): 143–167.

———. *Philosophers of Peace and War: Kant, Clausewitz, Marx, Engels and Tolstoy.* Cambridge: Cambridge University Press, 1978.

———. *Understanding War.* London: Routledge, 1991.

Who Was Who, 1996–2000, vol. 10. New York: Palgrave, 2001.

Peirce

Brent, Joseph. *Charles Sanders Peirce: A Life.* Bloomington: Indiana University Press, 1993.

Feibleman, James K. *An Introduction to Peirce's Philosophy Interpreted as a System.* New Orleans: Hauser Press, 1946.

Peirce, Charles Sanders. *Philosophical Writings of Peirce.* Edited by Justus Buchler. New York: Dover, 1955; first published 1940.

———. *Charles S. Peirce: Selected Writings (Values in a Universe of Chance).* Edited by Philip P. Wiener. New York: Dover, 1966; first published 1958.

———. *The Essential Peirce: Selected Philosophical Writings.* Vol. 1, *1867–1893.* Edited by Nathan Houser and Christian Kloesel. Bloomington: Indiana University Press, 1992.

Wittgenstein

Duffy, Bruce. *The World as I Found It.* New York: Ticknor and Fields, 1987.

Fann, K. T. *Wittgenstein's Conception of Philosophy.* Berkeley: University of California Press, 1971; first published 1969.

Hacker, P. M. S. "Wittgenstein, Ludwig Josef Johann," in H. C. G. Matthew and Brian Harrison, eds., *Oxford Dictionary of National Biography from the Earliest Times to the Year 2000.* Oxford: Oxford University Press, 2004.

Monk, Ray. *Ludwig Wittgenstein: The Duty of Genius.* New York: Free Press, 1990.

Wittgenstein, Ludwig. *Philosophical Investigations,* 3d ed. Translated by G. E. M. Anscombe. New York: Macmillan, 1971; first published 1953.

———. *Preliminary Studies for the "Philosophical Investigations" Generally Known as The Blue and Brown Books,* 2d ed. New York: Harper, 1965; first published 1958.

———. *On Certainty.* Translated by G. E. M. Anscombe and G. H. von Wright. Edited by Denis Paul and G. E. M. Anscombe. New York: Harper, 1972; first published 1969.

Collingwood

Browning, Gary K. *Rethinking R. G. Collingwood: Philosophy, Politics and the Unity of Theory and Practice.* London: Palgrave Macmillan, 2004.

Collingwood, Robin George. *Speculum Mentis, or The Map of Knowledge.* Oxford: Clarendon Press, 1924.

———. *Outlines of a Philosophy of Art.* Bristol, UK: Thoemmes Press, 1994; first published 1925.

———. *An Essay on Philosophical Method.* Bristol, UK: Thoemmes Press, 1995; first published 1933.

———. *An Autobiography.* Oxford: Oxford University Press, 1939.

———. *The New Leviathan, or, Man, Society, Civilization and Barbarism,* rev. ed. Edited and introduced by David Boucher. Oxford: Clarendon Press, 1992; first published 1942.

———. *The Idea of History.* Edited by T. M. Knox. Oxford: Oxford University Press, 1956; first published 1946.

———. *Essays in the Philosophy of History.* Edited by William Debbins. Austin: University of Texas Press, 1965.

———. *The Principles of History and Other Writings in Philosophy of History.* Edited by W. H. Dray and W. J. van der Dussen. Oxford: Oxford University Press, 1999.

Donagan, Alan. *The Later Philosophy of R. G. Collingwood.* Oxford: Clarendon Press, 1962.

Dray, William H. *History as Re-Enactment: R. G. Collingwood's Idea of History.* Oxford: Clarendon Press, 1995.

Johnson, Peter. *R. G. Collingwood: An Introduction.* Bristol, UK: Thoemmes Press, 1998.

Knox, T. M. "Collingwood, Robin George," in L. G. Wickham Legg and E. T. Williams, eds., *The Dictionary of National Biography, 1941–1950.* Oxford: Oxford University Press, 1959.

Martin, Rex. *Historical Explanation: Re-enactment and Practical Inference.* Ithaca, NY: Cornell University Press, 1977.

Tomlin, E. W. F. *R. G. Collingwood.* London: Longmans, Green, 1953.

Cognitive Science

Austin, James H. *Zen and the Brain: Toward an Understanding of Meditation and Consciousness.* Cambridge, MA: MIT Press, 1998.

Claxton, Guy. *Hare Brain, Tortoise Mind: How Intelligence Increases When You Think Less.* New York: Ecco, 1999; first published 1997.

Bibliography

Gardner, Howard. *The Mind's New Science: A History of the Cognitive Revolution.* New York: Basic Books, 1987.

Kandel, Eric R. *In Search of Memory: The Emergence of a New Science of Mind.* New York: W. W. Norton, 2006.

Kosko, Bart. *Fuzzy Thinking: The New Science of Fuzzy Logic.* New York: Hyperion, 1993.

Pinker, Steven. *How the Mind Works.* New York: W. W. Norton, 1997.

Music and Performance

Applegate, Celia. *Bach in Berlin: Nation and Culture in Mendelssohn's Revival of the* St. Matthew Passion. Ithaca, NY: Cornell University Press, 2005.

David, Hans T., and Arthur Mendel, eds. *The New Bach Reader: A Life of Johann Sebastian Bach in Letters and Documents.* Revised and enlarged by Christoph Wolf. New York: W. W. Norton, 1998.

Geck, Martin. *Johann Sebastian Bach: Life and Work.* Translated by John Hargraves. New York: Harcourt, 2006.

Taruskin, Richard. *Text and Act: Essays on Music and Performance.* Oxford: Oxford University Press, 1995.

Todd, R. Larry. *Mendelssohn: A Life in Music.* Oxford: Oxford University Press, 2003.

Wolff, Christoph. *Johann Sebastian Bach: The Learned Musician.* New York: W. W. Norton, 2000.

Studies on Alexander Pope

Fraser, George S. *Alexander Pope.* London: Routledge and Kegan Paul, 1978.

Mack, Maynard. *Collected in Himself: Essays Critical, Biographical, and Bibliographical on Pope and Some of His Contemporaries.* Newark, DE: University of Delaware Press, 1982.

———. *Alexander Pope: A Life.* New Haven, CT: Yale University Press, 1985.

Rogers, Pat, ed. *Alexander Pope.* Oxford: Oxford University Press, 1993.

Military Studies and Related Subjects

United States Army, Center of Military History. "The U.S. Army Chief of Staff's Professional Reading List." CMH Pub. 105-5-1 (2006).

United States Marine Corps. *Warfighting.* MCDP 1 (1997).

Index

Absolute war, xii, 4, 60, 67, 162, 168
  as abstraction, 123–124
  characteristics of, 70–71, 123
  conception of, 29, 73
  confusion about, 185
  defensive, 125, 126
  fighting, 71, 125
  genius and, 121–135
  goal of, 164
  guerrilla war and, 76, 186, 187
  offensive, 125, 126
  politics and, 31
  potential for, 169
  priority to, 168
  real war and, 70, 74, 75–76, 170–171, 185, 187
  reenactment and, 184
  theory and, 136
  violence and, 122–123
Action, xii, 56, 103, 117, 133, 178
  comprehensive guide to, 132
  delay in, 155, 172
  forms of, 160
  goals of, 188
  motives for, 154
  philosophy of, 67
  resorting to, 187
  suspension of, 156
  unilateral, 139
Adolphus, Gustavus, 95
Alexander I, 11, 83, 85, 96
Analysis
  historical, 45, 68, 80–94
  military, 13, 94, 195–196
  theoretical, 2, 45, 61, 138
  uncritical, 185, 195
  See also Critical analysis
Aristotle, 68
Armed civilians, 76, 84, 158, 159, 179
Armed conflict
  as center of gravity, 31, 32

friction and, 134–135
governing dynamics of, xvi, 1, 181
political nature of, xvi
practical understanding of, 3
study of, 121, 189, 192
theoretical significance of, 7
Army of the Rhine, 97
Aron, Raymond Claude Ferdinand, 56, 58, 59, 69, 70, 74
  Clausewitz and, 6, 37–50
  criticism of, 75
  Liddell Hart and, 40, 50
  on nonmilitary thought, 79
  On War and, xiii, 36, 37, 38, 39, 40, 42, 47, 48, 49, 50, 66, 77, 183
Art of War, The (Jomini), 16
Attack
  concentric, 156, 157
  defense and, 40, 121, 153–175
  merits of, 171, 191
Attackers
  motive of, 153, 166
  politics/policy of, 173, 186, 188
  strength of, 162, 180
  success for, 57
  weakening, 46, 47, 154, 166
Auerstädt, 82, 83, 88, 96
August, Prince, 97

Bach, Johann Sebastian, 197, 198
Bassford, Christopher, 7
Battle, importance of, 31–33, 161–162, 166, 180, 182
Battle of Hastings, 109
Bautzen, 86, 93, 96, 97
Bekenntnisdenkschrift (1812), 92
Beresina River, 89, 90, 96
Berlin, Isaiah, 1
Bernadotte, Jean-Baptiste Jules, 83
Beyerchen, Alan, 114, 115, 116
  advanced mathematics and, 112

Index

Beyerchen, Alan, (*continued*)
  linearization and, 113
  on manipulating reality, 113
  military theory and, 186
Blücher, Gebhard Leberecht von, 86, 87, 96
Bonaparte, Napoleon. *See* Napoleon
  Bonaparte
Borodino, 85, 96
Bourcet, Pierre de, 28
Boyen, Hermann von, 54
British Way in Warfare, The (Liddell Hart),
  28, 33, 35
Brodie, Bernard, xi, 76
Brunswick, Duke of, 82

Caesar, 14
Campaign of 1812 in Russia, The
  (Clausewitz), 87
Campaign of 1815 in France, The
  (Clausewitz), 87
Campaign of Trafalgar, The (Corbett), 21
Campo Formio, 155
Center of gravity, 31, 32, 173, 174
Chance, 102, 113, 114, 127–128
Chaos theory, xii, 77, 113
Clausewitz, Carl Philipp Gottfried von
  ambitions of, 97–98
  analytical vision of, 65
  combat experience of, 96
  conclusions about, 1, 7–8, 11–13
  death of, xiv, 2, 9, 191
  intellectual development of, 40–41, 52,
    95, 98
  marriage of, 78, 97
  method/style of, xvii, 95
  misunderstanding of, 29–30, 34, 49, 50,
    59, 72
  on *On War*, xiv
  original intention of, xiii, 9–10
  pedagogic/theoretical work of, 63–64
  as political theorist, 74
Clausewitz, Marie von, xiv, xv, 78, 97, 197,
  210n11
"Clausewitz, Nonlinearity, and the Unpredictability of War" (Beyerchen), 112
Clausewitz and the State (Paret), 36, 51, 60,
  63, 64

Clausewitz: Philosopher of War (Aron), 36,
  37, 40–41
Clausewitzian thought, 3, 45, 181, 186
  characterization of, 1, 5, 7, 42–43, 73, 115,
    176–181
  comprehension of, 182–183
  criticisms of, 41–42, 47
  psychological/historical genesis of, 52
  reductionist interpretations of, xiv
  teaching of, xii
Clausewitz Project, 51
"Clausewitz Today" (Gallie), 75
Claxton, Guy, 117, 118–119, 120
Code Napoléon, 153
Cognitive science, 77, 117
Collingwood, Robin George, 58, 76, 77, 101,
  104, 112, 118
  death of, 10
  education of, 106–107
  on experience, 109
  Gallie and, 65
  historical thinking and, 68
  on language, 107, 108
  reenactment and, 49, 109, 110, 184
  science of human affairs and, 111
  on war as continuation of policy, 209n61
Command, 19, 57
  psychology, 33
  strategic, 137
Commander-in-chief
  activity/boldness of, 16, 128
  adaptation by, 72
  decision-making by, xvi, xvii, 3, 23, 33,
    65, 67, 126, 127–128, 141, 145, 150, 169,
    177, 182, 195
  difficulties for, 43–44, 134
  effective, 24, 129, 143, 155
  emotions and, 128
  genius of, 3, 63, 170
  influence of, 169
  knowledge of, 132–133
  learning conditions and, 142
  psychological experience of, 4, 33, 131
  responsibility of, 127
  as statesman, 133
  truth and, 143
  will of, 35, 203n49

Concentration of force, 5, 21, 158, 174, 180
Conceptual system, 69, 72–73, 89
    effective, 182
    improving, 177
    intellect and, 144
    politics and, 178, 188–189
    study of, 182
Conscious mind, 135
Conscription, 25, 53, 84, 88
Convention of Kalisch (1813), 85
Convention of Tauroggen (1812), 85
Corbett, Julian
    Clausewitz and, 6, 17–25
    On War and, xi, 19, 24–25
Corn, Tony, xiii
Counterattacks, 17, 21, 22, 47, 93, 155, 157,
        163, 192
    defense and, 34, 58, 158
    effects of, 164
    opportunity for, 46
    overextension and, 89–90
    prelude to, 165
    retreat and, 90, 91
    time/space for, 15
Counteroffensives, 19, 34, 88, 190
Coup d'oeil, 130, 131, 170
Courage, xii, 47, 127, 128, 129, 131, 168
    intellect and, 130
Critical analysis, 100, 147, 148, 150, 151, 184
    historical knowledge and, 146
    importance of, 145–146
    model for, 197–198
    pictorial representation of, 195–196
Culminating point of victory, xiii, 167

Danchev, Alex, 35
Davout, Marshal Louis Nicholas, 82, 83
Decision-making, xvii, 3, 56, 67, 99, 101, 127–
        128, 131, 138, 142, 145, 148, 149, 167, 189
    conditioned, 100
    difficulties of, 175
    dynamics of, 135, 136, 151, 195
    effective, 117
    influences on, 14, 19, 45
    intuition and, 119
    nature of, 116
    political/policy factors on, 5, 182

powers of, 134
psychological circumstances of, 33, 34, 65
reenactment of, xvi, 45, 117, 136, 170, 178
strategic, xii, 14, 80, 96, 98, 112, 120, 129,
        130, 137, 169, 195
tactical, 141
understanding, 100
*Decisive Wars of History: A Study in Strat-
        egy, The* (Liddell Hart), 27, 34
Defence of Britain, The (Liddell Hart), 33
Defence of the West, The (Liddell Hart), 35
Defenders
    advantage for, 158, 159
    described, 160–161
    policy/politics of, 173, 186, 188
    will of, 153, 154, 162–163, 180
Defense
    attack and, 40, 121, 153–175
    character of, 157, 158–159, 163
    counterattacks and, 34, 58
    doctrine of strength of, 22
    essence of, 19, 21, 158, 159, 166
    failure of, 21, 164
    guerrilla war and, 48
    limited war and, 125
    mountain warfare and, 13
    naval, 21, 22
    offense and, 15, 18, 47, 76, 88, 91, 93, 94,
        155, 157, 159, 163, 164, 174, 203n52
    politics/policy and, 49, 178, 179
    strategic, 21, 46, 166, 181
    superiority of, 18, 33–34, 35, 46, 48, 57, 58,
        74, 91, 94, 128, 153–157, 160, 166, 171, 172,
        178, 179, 183, 186, 188, 190, 191, 192
    war and, 49, 158, 159
Determination, 130–131, 134
Diplomacy, 19
Doctrine, 98, 100, 152
    theory and, 39, 50
Drake, Francis, 17
Drake and the Tudor Navy (Corbett), 17

Emotion, xii, 3, 19, 109, 113, 115, 128, 132, 134,
        177
    accurate perception of, 105–106
    reenactment and, 178
    role of, 41, 44, 184

228

Index

Engagements, 141, 161, 182
  offered/refused, 203n48
  selective, xii–xiii, xiv
Engels, Friedrich, 65
England in the Mediterranean (Corbett), 18, 19, 20
England in the Seven Years' War (Corbett), 19, 20
Essay on Philosophical Method (Collingwood), 107
Eugene, Prince, 14
European Journal of Sociology, 66, 75
Experience, xvi, 94, 96, 109, 134–135, 153, 196
  genius and, 135
  moral, 140, 178
  as preceptor, 106
  psychological, 4, 33, 131
  synthetic, 184, 196
  theory and, 152

Fichte, Johann, 78
Firmness, 132, 134
First Cavalry Corps, 96
Fisch, Max H., 102
Fisher, John, 22
Fleming, Bruce, xiii
Foch, Ferdinand, 27, 28
Foch: The Man of Orleans (Liddell Hart), 27
Force
  concentration of, 5, 158, 174, 180
  dynamic play of, 159
  material, 139
  maximization of, 26, 121, 122, 124, 127
  moral, 133, 139, 140, 158
Frederick, 14, 149
Frederick William, Crown Prince, 95
Frederick William III, 9, 82–83, 84, 97
  defense and, 165
  military reorganization and, 83
  mobilization by, 81
  Napoleon and, 85
Free French, 37
French National Assembly, 80
French Revolution, 7, 11, 80, 157, 168, 190
Friction
  combat experience and, 134–135
  general friction and, 185

Fuller, J. F. C., 68
"Fundamental Principle of War, The" (Jomini), 16

Gallie, Walter Bryce (W. B.), 43, 49, 79, 111
  absolute war and, 74, 75, 76
  Clausewitz and, 6, 8, 64–77
  On War and, 36, 66, 67, 68–69, 71, 73–74, 75, 76, 183
  Peirce and, 101, 102, 103, 104
  on philosophy advances, 77
Gat, Azar, xiii, xiv, xv, 49, 78, 79
General friction, xvi, 134, 146, 177, 190
  friction and, 185
  negative effects of, 178
  play of, 147–148
Generalship, 176
  strategy and, 62
General War School, 78, 92
"Genesis of *On War*, The" (Paret), 63
Genius, 6, 43, 44, 61, 90, 102, 105, 155, 168
  absolute war and, 121–135
  action of, 150
  decision-making, 170
  elements of, 131, 132
  experience and, 135
  factual knowledge and, 149, 150
  nature of, 169
  quality of, 3, 44, 129
  role of, 3, 4, 63
  scientific investigation of, 139
German Movement, 79
German unification, Prussia and, 191
Ghost of Napoleon, The (Liddell Hart), 28, 30, 33
Glass Bead Game, The (Hesse), 193
Gneisenau, Augustus Wilhelm, 78, 86, 87, 92
  armed population and, 84
  Clausewitz and, 9, 97
Goethe, Johann Wolfgang von, 61
Gould, Stephen Jay, 8
Graham, J. J., 18, 66
Green Pamphlet (Corbett), 18, 19, 24
Grouchy, Marshal Marquis Emmanuel de, 87, 96
Guerrilla war, 4, 5, 35, 49, 51, 53, 74, 87, 88, 94, 124, 126, 173, 179, 182

absolute war and, 76, 186, 187
characterization of, 91, 187
countermeasures to, 92, 93
defense and, 48
development of, 91–92, 192
effectiveness of, 92, 164–165
real war and, 186, 187
resistance and, 188
resorting to, 165, 174, 186, 187, 190, 191
state-sponsored, 84, 125
theoretical, 164
violence of, 15–16
Guibert, Comte de, 28

*Hare Brain, Tortoise Mind: How Intelligence
    Increases When You Think Less*
    (Claxton), 117
Hegel, G. W. F., 39, 78, 197
dialectical form and, 5, 40
Herberg-Rothe, Andreas, 7
Hesse, Herman, 193
Heuser, Beatrice, 7
Historias, 197
Historical examples, 170
evaluation of, 44
proper/improper use of, 151–152
History, 108, 117, 133
philosophy of, 109
problems in, 45, 57
as record of human folly, 193
studying, xvi, 189, 194
theory and, xvi, 49, 57, 58, 59, 64, 76, 120,
    121, 135–153, 168, 170, 177–178, 182
Hobbes, Thomas, 74
Hohenlohe, Prince of, 82
Howard, Michael, 36, 66

Idea of History, The (Collingwood), 49, 65,
    111
Idealism, 60
realism and, 52
Ideas, 151, 152
Imperial Guard, 82
Improved capacity for judgment (ICJ), 196
Institute for Young Officers, 88
Intellect, 141, 148, 178
conduct of war and, 144

courage and, 130
importance of, 129, 132, 133
Intelligence, 119, 130, 132, 134, 141, 142
triumph for, 44
Intelligent unconscious, 118–119, 120
Intuition, 44, 118–119
decision-making and, 119
determination/fear and, 130
importance of, 183
improving, 3, 120, 135
nature of, 117
quality of, 133, 183, 187

James, William, 101
Jena, 82, 83
Jomini, Antoine Henri, 20, 23, 62–63, 97,
    177
Alexander and, 11
Clausewitz and, 5, 6, 11–14, 17, 208n44
concentration of force and, 180
criticism of, 186
on defensive-offensive, 15
facts/observations and, 63
on guerrilla war, 16
on military practice, 14
On War and, 17
on politics/strategy, 15

Kant, Immanuel, 37, 39, 40, 65, 78
principle of division and, 5, 70
Knowledge, 128, 132–133, 143
genius, 149, 150
historical, 146
self-, 101, 110
technical, 101

Language, 103–104, 110
communication and, 184
limitations of, 7, 98–100, 105, 106, 118, 183,
    184
objects/actions and, 118
proper use of, 99, 107–108
technical, 108, 111
Leadership, 22, 122, 190, 191
centers of gravity and, 174
irrational, 59
Leonard, Roger Ashley, 46

Index

Liddell Hart, Basil H., 68, 203n49
  Aron and, 40, 50
  Clausewitz and, 6, 25–35, 46, 50, 202n37
  criticism of, 29, 75
  on military history, 33
  On War and, 34
Ligny, 87, 96
Limited war, 4, 22, 41, 56, 63
  concept of, 23
  defense and, 125
  unlimited war and, 20, 47
  violence and, 60
Linearity, 112, 113, 114, 115, 186
Little wars, 92, 207n23
Locke, John, 74
Louise, Queen, 81
Ludendorff, Erich von, 27–28

Machiavelli, Niccolò, 1, 74, 78
Maneuver, 13–14, 26, 28, 72, 158, 163
Marlborough, Duke of, 14
Marx, Karl, 65
Material force, moral force and, 139
May, H. J., 18
Mendelssohn, Felix, 197
Military education, 55, 94, 115, 117, 120, 176,
    177, 181, 189
  importance of, 182
  theory and, 23
Military reform, 53, 54
Military Reorganization Commission, 83,
    84
Military Sciences for Young Infantry and
    Cavalry Officers, 78
Militias, 158, 159
Mind, 129, 134, 135, 140
Moltke, Helmuth von, xi
Monk, George, 17
Montaigne, Michel de, 78
Montesquieu, Baron de, 39, 40, 74, 78, 193
Moral dilemmas, 178, 180
Moral force, 133, 139, 140, 158
Moral values, 34, 35, 99–100, 112, 139, 140,
    156
Möser, Justus, 78
Mountain warfare, defense and, 13

Napoleon Bonaparte, 29, 51, 65, 88, 149, 155,
    169, 176
  absolute war and, 70
  on art of war, 15
  biography of, 11
  defeat of, 86, 87, 90
  guerrilla war and, 94
  injustices of, 54
  leadership of, 28, 80
  offense by, 84–85, 93, 158
  Prussia and, 80, 81, 82, 83, 86
  Prussian army and, 87
  wars against, 96
  withdrawal by, 85
Napoleonic Wars (1806–1815), 1, 78, 79, 88
Nero, 14
Newcastle, Duke of, 20, 21
Newton, Sir Isaac, 44, 169
Nonlinearity, 112, 113, 114, 116, 186–187
Nonpredictive theory, 103
Nonprescriptive strategic theory, 43
Novalis, 60

Observations on Prussia in Her Great Catas-
    trophe (Clausewitz), 87, 89
Occupation, 154, 181, 191
  problems of, 91
  resistance to, 5, 93
Offense
  advantages of, 21, 34
  call for, 175, 210n12
  consolidation of, 171
  defense and, 15, 18, 47, 76, 88, 91, 93, 94,
    155, 157, 159, 163, 164, 174, 203n52
  goal of, 164
  politics/policy and, 179, 191
  ruling out, 157
  strategic, 181
  tactical, 34
  weakening of, 165, 167
On War (Clausewitz)
  difficulties with, xii, xiii–xiv
  fragmented nature of, 188
  general meaning of, xii, 5, 6, 36
  influence of, xi, 1
  interpretation of, 6, 7, 181

231

Index

scrutiny of, xvii–xviii
theoretical perspective of, xiii, 2
translation of, 7, 36, 51, 63, 66
understanding, xvii, 52–53, 112, 121
as unfinished, xiv–xv, 1, 4, 49, 56–57, 66,
   69, 76
writing, xv, xvi, 98
Overextension, 46, 161, 176, 190
   counterattack and, 89–90

Pahlen, Peter, 96
Paret, Peter, 43, 69, 70, 73, 74, 93
   absolute/real war and, 75
   Clausewitz and, 6, 50–64, 79
   criticism of, 59, 75
   little wars and, 207n23
   method/style of, 95
   On War and, xi, 51, 52, 55, 56, 57, 59, 61,
      63, 66
   on Scharnhorst, 95
   on study of war, 94
   translation by, 36
Paris, or the Future of War (Liddell Hart),
   25
Paris Convention (1808), 84
Passion, 47, 124, 127, 141
Passion (Music), 197
Peirce, Benjamin, 101
Peirce, Charles Sanders, 76, 77, 101, 105, 107,
   111, 112, 118, 183, 208n53
   Clausewitz and, 102, 104
   Gallie and, 64, 65, 68, 102, 103
Peirce and Pragmatism (Gallie), 64, 102
Peninsular War, 91
People's War. *See* Guerrilla war
Personal enlightenment, 193
Pestalozzi, Johann Heinrich, 78
Phenomena, 112, 186
   nature/dynamics of, 118
   theory of, 2, 185
Philosophers of Peace and War (Gallie), 36,
   66, 75
Philosophical invention, 94–112
Philosophy and the Historical Understanding
   (Gallie), 65
Phull, Karl Ludwig von, 90, 96

Pitt, William, 20
Pitt's War, 20
Plato, 65
Policy, 48, 173
   politics and, 178
   strategy and, 133
   war and, 42, 47
   war as continuation of/by other means,
      xii, xiv, 4, 18, 22, 172, 180, 186, 209n61
Political/policy motives, 179
   achieving, 172
   extension of, 187
   military action and, xii
   modification of, 192
   strategy and, 162
Politics, 51, 58, 122, 176
   absolute war and, 31
   conduct of war and, 188–189
   decision-making and, 5
   defense and, 49
   policy and, 178
   strategy and, 15, 19, 97
   war and, xv, 14, 31, 48, 49, 50, 55–56, 57,
      59, 63, 73, 74, 115, 124, 127, 162, 163, 169,
      172, 173, 180, 187, 190, 191
Pope, Alexander: verse by, 192–193
Popular insurrection, 154, 164
Positive doctrine, 140–141
Powell, Colin, xi
Practice, theory and, 44, 138, 185
Predictive theory, 102, 103, 106
Prescriptive theory, 41
Principle of division, 5, 70
Principles (of war), 15–17, 20, 144, 152–153
   (general propositions), 170, 180, 183,
      188, 195 (general propositions)
   theory and, 141
Problem-solving, 60, 65, 94, 100, 110, 120
Protracted war, 35, 47, 124, 162
   will for, 180–181
Prussia
   defeat of, 53, 80, 83
   German unification and, 191
   inferiority of, 176
   moral mission of, 53
   transformation of, 191–192

Index

Prussian army, 53, 54, 93
  changes for, 83, 84
  command of, 86
  deployment of, 81
  return of, 82, 190
Psychological experiment, 136
Psychological factors, 62, 120, 136–137

Ramsey, Frank P., 105
Rapoport, Anatol, 46
Rationality, xiv, 47, 113
Realism, idealism and, 52
Reality
  recreating, 113, 178
  theory and, 45, 62, 95, 144
Real war, xii, 4, 43, 125, 162, 168
  absolute war and, 70, 74, 75–76, 170–171,
    185, 187
  guerrilla war and, 186, 187
  potential for, 169
  reenactment and, 184
  social function of, 70
  theory and, 136
Real War, 1914–1918, The (Liddell Hart), 27
Reason, xii, 120, 127, 137, 184
Reenactment, xvi, 4, 68, 94, 110, 111, 117, 127,
    145, 146, 152, 153, 170, 179, 183, 189–190,
    197–198
  absolute war and, 184
  applying, 120, 190
  authentic, 149, 178
  concept of, 109, 185
  decision-making and, 136, 178
  drawbacks of, 177
  emotions and, 178
  historical, 49, 65, 177, 184
  objective of, 45
  procedural, 149
  psychological, 101
  real war and, 184
Reflection on synthetic experience (RSE), 196
Reflection on verifiable fact (RVHF), 195
Remaking of Modern Armies, The (Liddell
    Hart), 26–27
Resistance, 74, 91, 156, 160, 163, 164, 165
  effectiveness of, 155, 179
  guerrilla war and, 188

  moral, 154
  overcoming, 131
  period of, 155
  physical, 154
  political support for, 179
  popular, 174, 190
Responsibility, 127, 131, 137, 167
Retreat, 154
  buying time with, 46, 179
  counterattacks and, 90, 91
  overextension and, 90
  psychological effects of, 164
  voluntary, 163–164
Rosinski, Herbert, xi, 37
Rousseau, Jean Jacques, 74
Routine, use of, 144–145
Royal Naval College, 18
Royce, Josiah, 101
Russian army, 53, 85
Russian campaign (1812), 163–164, 171

Saxe, Marshall, 28
Scharnhorst, Gerhard Johann David von,
    41, 84, 89, 91, 97
  Clausewitz and, 88, 96
  death of, 86
  defense and, 88
  guerrilla war and, 92
  influence of, 94
  Military Reorganization Commission
    and, 83
  theory/reality and, 95
Schaumburg-Lippe-Bückeburg, Count
    Friedrich Wilhelm Ernst zu, 88
Schiller, Friedrich, 61
Schlegel brothers, 60
Schlieffen, Alfred von, 27
Scientific perspective, 111, 112–120, 183
Scipio Africanus, 14, 26
Sea power, state power and, 20
Security, 176, 191
Seven Years' War, 20, 21
Siege warfare, 138
Slade, Edmond John Warre, 18
Smolensk, 85, 90
Some Principles of Maritime Strategy (Cor-
    bett), 22, 24